KU-185-531

WHEN THE
EARTH
WAS
FLAT

By the same author

The Long and the Short of It
Learn on the Loo
On This Day in History
The Accidental Scientist

ALL THE BITS OF SCIENCE WE GOT WRONG

GRAEME DONALD

Michael O'Mara Books Limited

For Rhona, as ever, and with very special thanks to Kath 'Kay-Dee' Davies, who selflessly moved heaven and earth to give me the time to bring the manuscript in on time.

This paperback edition first published in 2017

First published in Great Britain in 2012 by
Michael O'Mara Books Limited
9 Lion Yard
Tremadoc Road
London SW4 7NQ

Copyright © Michael O'Mara Books Limited 2012, 2017

All rights reserved. You may not copy, store, distribute, transmit, reproduce or otherwise make available this publication (or any part of it) in any form, or by any means (electronic, digital, optical, mechanical, photocopying, recording or otherwise), without the prior written permission of the publisher. Any person who does any unauthorized act in relation to this publication may be liable to criminal prosecution and civil claims for damages.

A CIP catalogue record for this book is available from the British Library.

Papers used by Michael O'Mara Books Limited are natural, recyclable products made from wood grown in sustainable forests. The manufacturing processes conform to the environmental regulations of the country of origin.

ISBN: 978-1-78243-783-3 in paperback print format
ISBN: 978-1-84317-925-2 in Epub format
ISBN: 978-1-84317-926-9 in Mobipocket format

Designed and typeset by K DESIGN, Somerset

Printed and bound in Great Britain by CPI Group (UK) Ltd, Croydon
CR0 4YY

www.mombooks.com

FSC
www.fsc.org
MIX
Paper from responsible sources
FSC® C013604

Contents

Introduction

FROM ANCIENT TIMES to the modern day, science has strayed many times from the truth.

Often these discoveries have been dictated by the constraints of contemporary thought. Limited by their lack of knowledge of the human anatomy, the Ancient Greeks developed the theory that the body is made up of four humours, an idea that held sway until the march of scientific medicine in the nineteenth century.

Other times these ideas were the result of pure folly, such as the seemingly innocuous development of phrenology – a concept used to justify genocide in Rwanda in the late twentieth century. Sometimes scientific 'facts' have been explored in an effort to lend false support to a hidden agendum, including the appropriation of the entirely spurious notion of subliminal messaging by politicians and the Christian Right. Despite the questionable nature of these ideas, *When the Earth Was Flat* will highlight how man has been – and perhaps always will be – at the mercy of science.

Thankfully not all the bits of science we got wrong have had such a devastating impact; some of the examples in this book will raise a smile. Whether it's the alchemists' search for the philosopher's stone – the vehicle through which all base metals could be turned into gold – the somewhat surprising history of the vibrator, or the many proponents of the hollow earth theory, the annals of science are littered with strange people and their even stranger ideas.

What is perhaps most surprising is that some of science's most spurious ideas have only recently been relinquished. No matter how advanced today's medical and scientific thinking might be, who is to say that in one hundred years' time a book similar to this one won't be ridiculing today's received wisdom.

Having Your Bumps Felt

The physical measurements of the skull correlate to a person's personality

THE MAJORITY OF the scientific frivolities of previous centuries inflicted little or no real harm during their reign, and evaporated without much trace in the light of new discoveries. Unfortunately, the same cannot be said for the pseudoscience of phrenology, which caused wide-ranging injustices and misery in its time and, most damaging of all, reached out from its own grave to promote genocide at the close of the twentieth century.

THE GALL OF IT

The father of phrenology was the German physician Franz Josef Gall (1758–1828), a product of the University of Vienna, an institution that served as a breeding ground for several other spurious notions about the human race (see box on page 11). Gall developed the theory that the human brain is comprised of twenty-seven distinct zones, each of which is a wholly separate and autonomous organ with

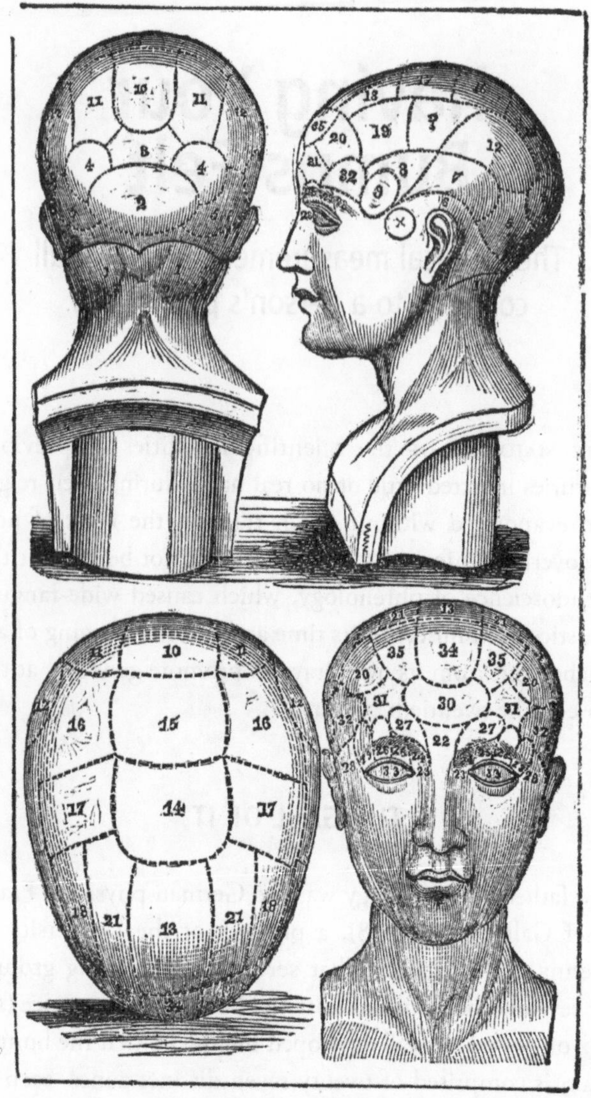

The phrenological bust

individual responsibility for certain functions, characteristics and predispositions.

LESSONS IN IDIOCY

By 1925 the University of Vienna had become an intellectual hotbed of racist ideology. The most notorious and far-reaching of such notions was *Rassenpflege* – the quest for racial hygiene. Professor Otto Reche, director of the university's Department of Anthropology, was the most vocal proponent of such ideas, proclaiming, '*Rassenpflege* must be the basis for all domestic policy and at least a part of foreign policy as well.'

The more an individual used one of the zones, or allowed themselves to be driven by the emotional or physical urges dictated by it, the larger that zone would become – similar to an overused muscle. In Gall's defence, his findings were not completely off the mark: it is now known that certain areas of the brain are linked to specific functions or temperament, and that some of these areas can become enlarged with mental exercise.

Had Franz Gall finalized his research at this point, no harm would have resulted. His error was in expanding the basic premise into the foundation stone of a sizeable edifice of speculation and assumption. By 1805 Gall had decided that the twenty-seven zones must be responsible for the

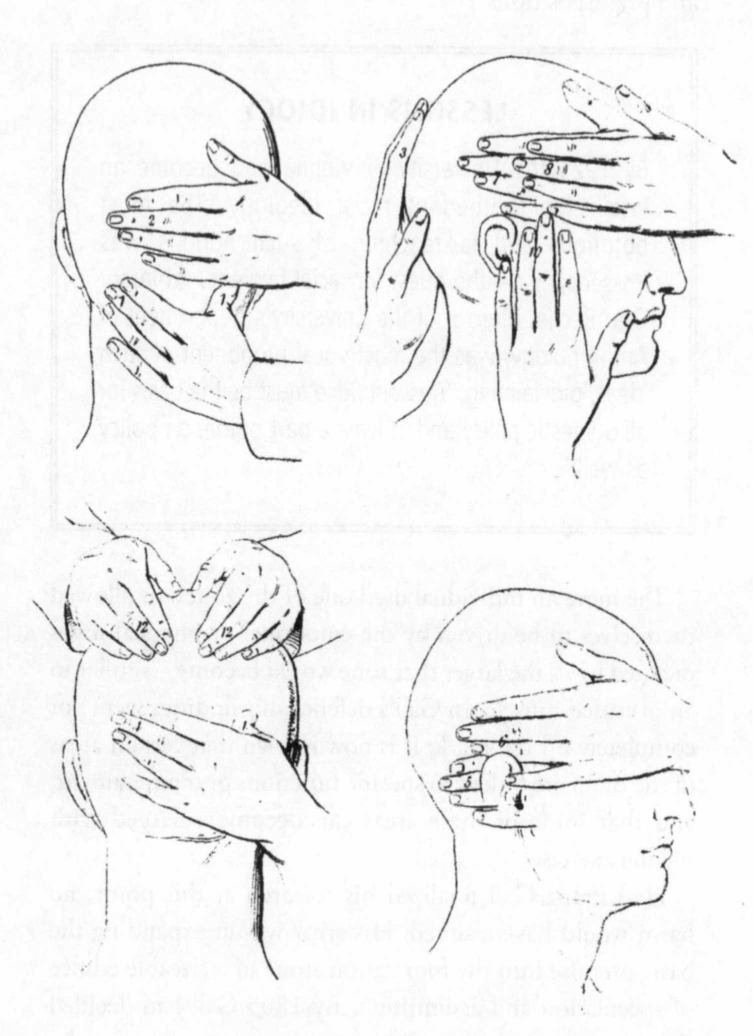

Someone having their bumps felt

lumps and bumps on the anterior of the skull, which they pressed against as they swelled with exertion.

BRAIN TRAINING

In March 2000, Professor Eleanor Maguire of University College London published the results of an extensive study she had conducted into the pattern of growth of the hippocampi in the brains of London taxi drivers. They were chosen because they are required to take 'the knowledge', the formidable exam that demonstrates their ability to work out the best route between any two nominated points in the city. Professor Maguire deduced that the longer the driver had been working, the more pronounced the enlargement of the hippocampus.

LUNACY

Gall conducted exhaustive fingertip exploration on the skulls of murderers, burglars and other categories of criminals and decided that there were sufficient significant similarities between them to establish a pattern. He also conducted similar explorations on the skulls of the insane and decided that their individual conditions were attributable to specific zonal malfunction. Again, in defence of Gall, some good did come of this thinking, as the insane were previously thought to be wilfully so or possessed by the Devil, and thus were

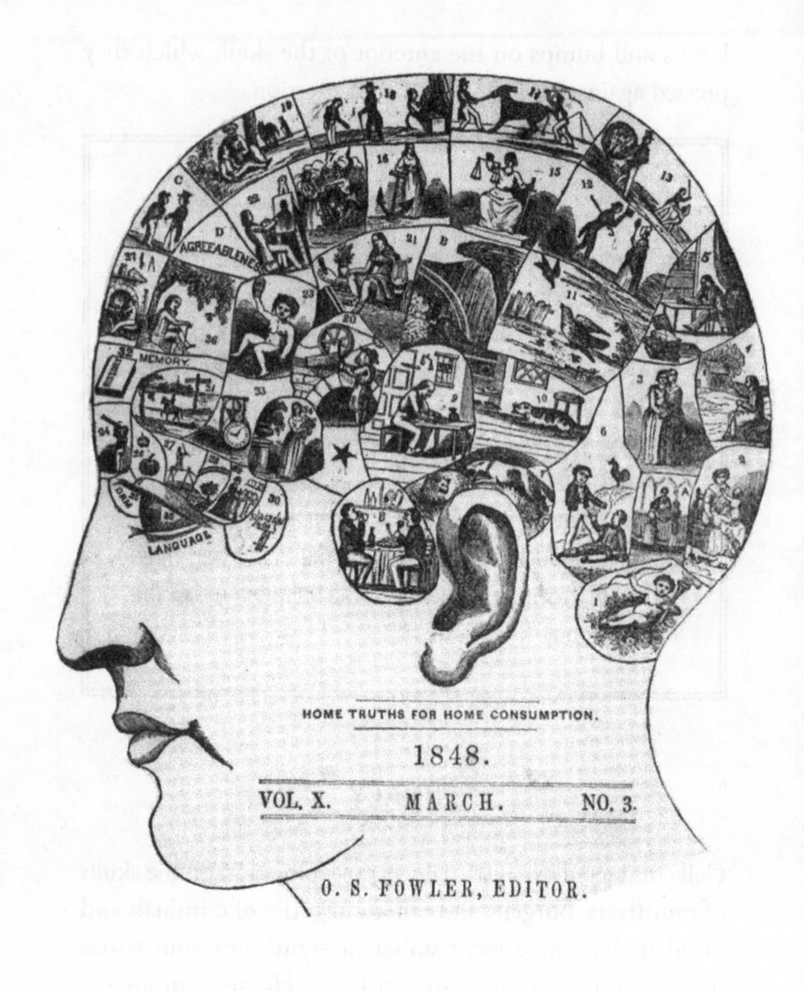

From the March 1848 edition of the
American Phrenological Journal, edited by Orson Fowler

beaten regularly. Such was the standing of phrenology that, almost overnight, the insane were for the first time regarded as genuinely ill and treated accordingly.

But this did not bode well for other people who, despite having previously led perfectly normal lives, happened to possess a few lumps and bumps similar to Gall's 'scientifically proven' pattern, marking them out as potential murderers or lunatics. A few unfortunate souls found themselves locked up as a preventative measure. The public too became seduced by Gall's theories, which they picked up second-hand from the writings of celebrity converts including the Brontë sisters, Bram Stoker and, most popular of all, the Sherlock Holmes yarns of Conan Doyle. If it was good enough for Holmes then it *had* to be right.

Companies incorporated phrenology in their personnel selection, with 'experts' fondling the heads of prospective employees to ensure their clients were not about to hire a lunatic. In the courts, many defendants were imprisoned through convictions secured, in part, on the 'expert witness' ramblings of professional phrenologists. But the cracks in Gall's edifice were already apparent by 1820 and by 1850 it lay in ruins – but only in the UK.

COMMITTING A FOWL

Phrenology was by this time deeply entrenched in the United States, in the main through the efforts of the Fowler brothers, Orson (1809–87) and Lorenzo (1811–96), who counted the likes of the American essayist Ralph Waldo

Emerson (1803–82) and the inventor Thomas Edison (1847–1931) among their supporters. It would be unkind to brand the Fowlers as complete charlatans, but it should be acknowledged that both had an eye for a fast buck, especially Lorenzo, who visited the UK in 1860 for a lecture tour that proved so lucrative he decided to stay.

While in London Lorenzo established the Fowler Institute where, in 1872, the American author and humorist Mark Twain tried in vain to expose him. An inveterate prankster, Twain donned a lower-middle-class disguise and booked a reading during which Fowler, who showed little interest in his subject beyond the collection of the fee, identified a significant depression in Twain's skull, which, he claimed, indicated a total lack of any sense of humour. The subject was also, in Fowler's professional opinion, lacking in any creative ability and best suited to mundane work of a clerical nature. Twain mumbled his humble thanks, paid up and left.

IT FELT LIKE A GOOD IDEA AT THE TIME

In 1958 Dr Edmund Teller (the 'father of the H-bomb') proposed detonating a chain of his 'children' to create a mile-wide harbour at Cape Thompson in Alaska. Thankfully, the idea was eventually shelved.

A month or so later, Twain rebooked under his own name and turned up in his trademark white suit, full of bravado and swagger. This time a much more obsequious Fowler lionized his celebrity client and, at the exact same spot on Twain's skull that had on the previous meeting presented a depression, there now appeared a 'mountainous protrusion' consistent with the star's international reputation as a humorist. Twain paid his fee and left to publish the results. But nothing could stop the Fowler bandwagon. Lorenzo had by then set up a sizeable mail-order operation to provide the burgeoning craze for phrenology parties with all the necessary paraphernalia.

Any one of the iconic beige-coloured phrenology busts complete with black markings (as seen in antique shops today) is likely to be one of Lorenzo's products. All harmless fun, perhaps, and certainly no more dangerous than the craze for ouija sessions that was to follow. Fowler's only other legacy has been the introduction into the English lexicon of expressions such as 'high-brow', 'low-brow', and anyone acting irrationally being told it is time they 'had their bumps felt'. But things were about to get worse – much worse.

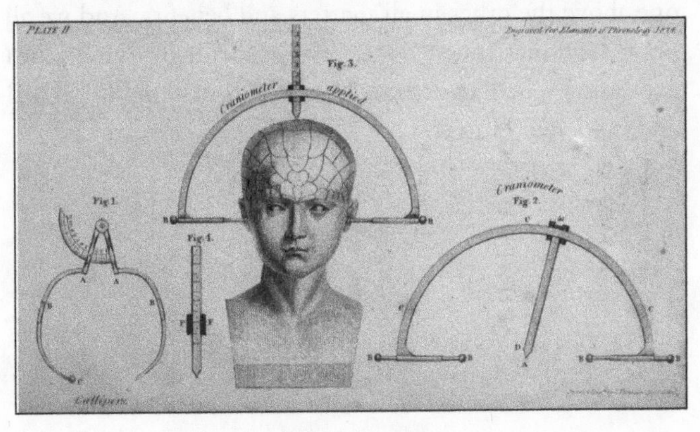

Measuring up

A TURN FOR THE WORSE

The Treaty of Versailles (1919) turned the former German colony of Rwanda over to Belgian control. Proceedings took a dark turn when Belgium too fell foul to the phrenology craze, under the guiding hand of its leading proponent, Paul Bouts (1900–90). A priest-phrenologist who by the age of twenty-four was already a Belgian national celebrity, Bouts visited a variety of institutions throughout his homeland and measured inmates' heads with the aid of his self-designed instruments. He used his findings to make dubious pronouncements on who was 'normal' and who was not.

To make matters worse, racial overtones began to creep in when Bouts' devices were used by the Belgian Colonial Office in Rwanda to decide on matters of racial superiority. After clamping a few heads in their mail-order callipers, the office pronounced the Tutsi to be racially superior to the Hutu, and treated the two accordingly, setting the one above the other in all matters and benefits. And we all know how that ended – the 1994 genocide in which Hutu extremists killed an estimated 500,000 to 1 million Tutsi and moderate Hutu.

Bad Vibrations

A battalion of marching soldiers can cause a suspension bridge to collapse

IN THE NINETEENTH century the military was warned that all bodies of marching soldiers, from a single platoon to a full regiment, should always break step when passing over a bridge. The advice was reinforced by contemporary scientific discussions about the way in which all objects possess a natural frequency – the frequency with which something will vibrate once it has been set in motion. It was believed that if the repeated and synchronized step of the soldiers marching in cadence matched the natural frequency of the bridge they were crossing, catastrophe would inevitably ensue.

TROUBLED WATER

The notion was born of the Broughton Suspension Bridge disaster of 12 April 1831. Built in 1826 at the personal expense of wealthy Mancunian John Fitzgerald, the bridge spanned the River Irwell between Broughton and Pendleton in Lancashire. On the day in question, Lieutenant John

Fitzgerald Jnr. was leading seventy-four members of the 60th Rifle Corps back from exercise on the moors to the barracks in Salford. As they crossed the bridge, marching proudly in step, the structure began to collapse and the entire column of men was tipped into the river. Fortunately the water was only approximately half a metre deep and any injuries suffered were minor.

Scientists were immediately consulted at the newly opened Manchester Mechanics' Institute, to which Fitzgerald had made considerable donations. They deduced that the collapse had been brought about by the resonance caused by the soldiers stamping along in unison. This conclusion brought some comfort to those who had invested in the suspension bridge – one of the earliest of its kind, the Broughton span was a matter of local pride and those who

WELL I NEVER! POPULAR SCIENTIFIC IDEAS DEBUNKED

- There is no such thing as a centrifugal force.
- Heat does not rise but disperses itself equally and evenly throughout its environment.
- Stomach ulcers are not caused by stress or by spicy food but by a bacterium called *Helicobacter pylori*.
- A quantum leap does not denote a seismic step in progress, but a minute transitional change that occurs when matter shifts from one state to another without any discernible change in the process.

had designed and built it did not wish to find themselves accused of incompetence, or worse. The military duly sent out an immediate instruction to all marching units, big or small, that soldiers were to break step and walk casually across any bridge for fear of bringing it crashing down.

NUTS AND BOLTS

In reality, mechanical resonance, although a very real force, had nothing to do with the issue, and nor had the marching troops. When the fuss died down and engineers unconnected to the bridge's patron investigated the site they found that one of the large bolts that secured one of the stay-chains to the ground anchor had snapped. It was also discovered that many of the other bolts that anchored the suspending chains were either cracked or bent, and the bolts used were three-year-old replacements of bolts that had failed before.

More pointed questioning revealed that the pre-eminent structural engineer Eaton Hodgkinson (1789–1861) had expressed doubts about the strength of the chains and advised they be tested before being installed on site; wise words that went unheeded. Additionally, if the soldiers' marching in step was the unforeseen harbinger of the bridge's doom, why had it not collapsed when the troops had marched over it on their way out to Kersal Moor? In effect, the bridge was *ready* to fall down and it just happened to do so under the weight of the Corps; marching in step had nothing to do with the matter. It had been simple mechanical failure because the bridge had been badly designed and built.

THE MYTH TAKES HOLD

Yet the myth that synchronized heavy footing could cause a bridge to fall to pieces continued and was further reinforced by the collapse of the Angers Suspension Bridge in France on 16 April 1850. The bridge collapsed after two suspension cables snapped when a battalion of some 500 troops marched across it amid a violent thunderstorm. A total of 226 soldiers lost their lives. Yet again marching-induced mechanical resonance was blamed, despite the fact the soldiers had been ordered to double space and break step. In addition there was a significant troop presence in the area and whole battalions routinely used the bridge, some breaking step and others not. On 16 April two battalions from the same regiment had crossed the bridge earlier in the day, without incident. Yet again corrosion problems were found at the anchor-points of the snapped cables. As in the case of the Broughton suspension bridge, the collapse of the Angers bridge was the result of a simple mechanical failure.

The collapse of the Angers suspension bridge

MILLENNIUM WOBBLE

In an article titled 'London Bridge's Wobble and Sway' published in *Physics Today* in March 2010, physics professor Bernard J. Feldman challenged the argument that the wobble experienced on London's newly opened Millennium Bridge in June 2000 was the result of synchronized resonance. Key to his claim was that the walking frequency of pedestrians is double the lateral oscillation of bridges, and thus unlikely to have any impact.

HIGH WINDS

The spectacular collapse of the Tacoma Narrows Suspension Bridge over America's Puget Sound in 1940 was automatically assumed to have been caused by wind-induced resonance. The bridge had already earned itself the nickname 'Galloping Gertie' due to the way the deck reared about, even during construction. Despite this, the cataclysmic nature of the bridge's later collapse only ran to one casualty: a spaniel called Tubby.

Although the bridge was supposedly built to withstand 120mph winds, the disaster happened in 40mph winds. Nevertheless, wind-induced resonance was immediately blamed. The wind rushing past the bridge was thought to have created a stream of whirlwinds, the fluctuations of which matched the bridge's natural frequency. The vibrations then reached such a pitch the bridge was forced to collapse.

THE BACKLASH BEGINS

Today, talk of Gertie falling foul of mechanical or wind-induced resonance still abounds, but with a few exceptions. Robert H. Scanlon (1914–2001) wrote several papers that lambasted this misconception and, as the main consultant on the Golden Gate Bridge project, his comments came with some authority. Hailed internationally as the 'father' of the study of the aerodynamics and aeroelasticity of such structures, Scanlon, along with other leading lights in the field, repeatedly poured cold water on the Tacoma resonance theory.

Professor P. Joseph McKenna and Professor Alan C. Lazer's article 'Rock and Roll Bridge' leads a very convincing case against the Tacoma resonance theory. For them, resonance is a very precise entity. Using the shattering of glass as an example, McKenna and Lazer describe the unique circumstances needed for the forcing frequency to match the natural frequency of the object. Such 'precise, steady conditions' are unlikely to have been in place during the powerful storm that hit the Tacoma Bridge. Rather, they attribute the bridge's demise to the different types of oscillation it experienced during the storm, which resulted in an extreme twisting of the roadway. I might also add that the pressure on the suspension cables as the Tacoma carriageway lifted and fell violently in the wind would not have helped matters either.

Despite the unique circumstances of their individual demise, the collapsing of the Broughton, Angers and Tacoma bridges had nothing to do with resonance theory. Yet the old errors of science do not always die that easily, which is why to this day troops will always break step on bridges, just in case there is anything in the old wives' tale.

HITTING A HIGH NOTE

To the other great resonance myth: glass can be shattered by the human voice. Scientists of the nineteenth century believed an opera singer could hit and hold a note long enough to shatter a drinking glass. Despite the number of ear-splitting salon demonstrations, something was amiss – the human voice simply is not powerful enough to shatter glass. But whichever trick was used – an accomplice with an air-pistol would not have been heard amidst the racket – science was once again duped into extolling the power of resonance.

More recently, a famous television advertisement showed Ella Fitzgerald apparently pulling off a similar trick, but that too was rigged. The glass itself holds the secret: it has first to be 'pinged' to reveal its own note of resonance; this must then be recorded and played back in the direction of the glass through loudspeakers until the glass shatters. The human voice lacks the necessary power – volume holds the key.

Going For Gold

All base metals can be turned into gold

The origin of alchemy, the father of modern chemistry, while obscure, is thought by some to have derived from the Arabic *al-Khemia*, or the (land of the) Black Earth, an ancient epithet of Egypt. Europe's introduction to alchemy occurred during the eleventh century, when it was introduced to Spain by the Moors. Aside from a few grand principles, which included plans to discover the secrets to eternal life, alchemy's prime goal (and the one for which the 'science' is best remembered) was the search for the 'philosopher's stone' – the vehicle through which base metals could be turned into gold.

ELEMENTARY MATTERS

The basis of alchemy was the Aristotelian concept that all matter was alike – a cabbage and a brick comprised exactly the same substances, they just assumed a different form and inspirational spirit. In order to turn a cabbage into a brick, for example – or a lump of lead into gold – a person needed to first identify the 'spirit' of the cabbage or the brick and imbue the one with the other.

Although alchemists acknowledged the four classical elements of earth, fire, water and air, they solidly regarded them as different manifestations of the same, singular matter. If a person were to heat water, for example, it would become air; if the air was chilled, water appeared: natural phenomena seen by the alchemists as a validation of their basic premise.

In alchemists' circles, the search for the philosopher's stone was termed the Magnum Opus, a tag now applied to a person's great labour. But the question that remains unanswered is: why were so many bright people duped into believing such an absurd principle? If lead could be turned so easily into gold then the price would drop out of the gold market and gold would become as cheap as, well, lead. But greed, it seems, blinded all – medieval Europe teemed with charlatan alchemists who happily ripped off well-heeled and avaricious nobles, all too eager to part with their money after witnessing a few shabby tricks dressed up to look like miracles.

An artistic representation of the philosopher's stone

CELEBRITY CONVERTS

Not all alchemists were focused exclusively on fleecing the gullible; some very bright minds joined the quest for the key to all matter and made significant contributions to science and medicine along the way. The alchemist Paracelsus (1493–1541) was the first to identify and name zinc. He also invented laudanum, an alcoholic solution that contained morphine, later guzzled enthusiastically by Victorian English matriarchs until over-the-counter sales of opium products became illegal in 1920 (see Victoria's Secret on page 94).

WELL I NEVER! POPULAR SCIENTIFIC IDEAS DEBUNKED

- There is no such thing as a tongue map: sweet, sour, salt and other flavours can be detected all over the organ.
- It is your sense of smell that you lose with a cold, not your sense of taste.
- 'Sixth sense' is a silly expression because humans actually possess nineteen senses.

The lure of alchemy was strong and in Paracelsus' time, and for one or two centuries afterwards, the dividing line between the more ethical strands of alchemy and mainstream science was blurred, with John Dee (1527–*c.*1608), consultant to Queen Elizabeth I, and Sir Isaac Newton

An image from Michael Maier's alchemical emblem book *Atalanta fugiens.* Gold and silver (the sun and the moon) are shown to be in conjunction

(1642–1727), one of the science world's most influential pioneers, both dabbling on the dark side. But not all such experimentations ended well. The celebrated alchemist Dr Faustus (1480–1540), having made countless enemies in the ranks of the clergy, a few of whom he managed to poison with his alchemic remedies, blew himself to smithereens while he experimented with glycerine and acids in his search for the 'Water of Life'.

If the acid Dr Faustus had used was nitric acid then it is little wonder there was nothing left of him: he may have been 200 years ahead of the invention of nitroglycerine, a highly explosive liquid which has since been used in the production of dynamite. Regardless, the Church explained the absence of bodily remains as the result of the Devil's work.

RICH EXPERIMENTATION

If Faustus had discovered nitroglycerine, if only for a fraction of the second he took to dispatch himself, he was not the only alchemist to be ahead of the march of traditional science. But the sinister reputation of the art clouded alchemists' discoveries with suspicion and papal objections, subsequently holding back scientific development. The basic principle held that if it came out of an alchemist's workshop then it was likely the work of the Devil himself.

One such pioneer was the Polish alchemist Michael Sendivogios (1566–1636), who produced oxygen by heating nitres almost 200 years before theologian Joseph Priestly (1733–1804) was acclaimed for the 'discovery' of the same in 1774. Sendivogios successfully managed to share his knowledge with the Dutch alchemist Cornelis Drebbel (1572–1633), who put it to great and advanced practical use. In London in 1620, Drebbel built the first dirigible submarine, capable of carrying sixteen people.

Drebbel had discovered that by burning potassium nitrate or sodium nitrate he could not only produce oxygen, but the process also transmuted the nitrates to an oxide or hydroxide that absorbed the build-up of carbon dioxide. Thus, he too found himself 300 years ahead of his time by producing a crude but effective re-breathing system. The craft was tested with a

Heading into the laboratory: (a) copper still; (b) still head; (c) cooling medium; (d) condensing tube; (e) receiver

full crew in the Thames in front of James I and the Navy. It stayed submerged for over three hours as it travelled up and down the river at a depth of approximately five metres. But again whispers of satanic involvement abounded, and the Navy was deprived of functioning submarines for war purposes.

BAD REPUTATION

Despite such pioneering discoveries, it was the charlatans who grabbed the limelight and dragged the name of alchemy through the mud. The court of the Habsburgs found itself the most vulnerable to alchemy's dark side. Holy Roman

Emperor Ferdinand III (1608–58) was duped into believing he had witnessed the creation of a nugget of gold; he heaped a fortune on the Austrian alchemist Johann Richthausen, who promptly made off with the proceeds. Leopold I (1608–58) was similarly hoodwinked, and it fell to the wiser Empress Maria Theresa (1717–80), the last of the House of Habsburg, to ban all attempts at transmutation throughout her realm.

But now it seems that the worst of alchemy's breed may have been on to something after all. Today, particle accelerators, like the Large Hadron Collider on the Franco-Swiss border, routinely transmute a variety of elements by knocking free neutrons and protons from one element, or by bombarding that same element with protons from another. So, while transmutation by chemical means may well be impossible, it is not so in the realm of physics. In 1972 Soviet physicists at the research facility on the shores of Lake Baikal in Siberia reported that, on a routine inspection, the lead lining of the deflector shields in an experimental reactor had turned to gold. Naturally, this was questioned by a sceptical West until Glenn Seaborg, Nobel Laureate for Chemistry, managed to achieve the same result at the University of California in 1980.

Under the arm of nuclear physics, Seaborg successfully transformed several thousand atoms of lead and bismuth into gold by removing certain neutrons and protons from the sample. Although this may go some way to vindicate the notions of the early alchemists, the sheer expense of the operation results in gold costing many thousand times that which is mined in the conventional way, so the gold market may rest easy for some time yet.

Good Vibrations

Hysteria is the sole preserve of women and it can only be alleviated by genital stimulation

An etymological sister of the word hysterectomy, hysteria is derived from the Greek *hustera*, meaning womb. The condition was, from ancient times and until quite recently, believed in medical circles to be the exclusive province of women, and caused by an imbalance in their wombs or vaginas; a somewhat far-fetched notion rendered more farcical by the suggested treatment for the condition that constituted accepted medical practice until well into the twentieth century. While such treatments would today have its practitioners immediately struck off the medical register, their acceptance in the annals of medicine led indirectly to the development of the vibrating sex-toy industry.

PAROXYSMS OF JOY

In 1563 the Dutch physician Pieter van Foreest (1521–97) concurred with the centuries-old remedy for 'hysteria', or 'womb disease', when he wrote the following in his published collection of medical observations:

> When these symptoms indicate, we think it necessary to ask a midwife to assist, so that she can massage the genitalia with one finger inside, using oil of lilies, musk root, crocus, or similar. And in this way the afflicted woman can be aroused to the paroxysm. This kind of stimulation with the finger is recommended by Galen and Avicenna, among others, most especially for widows, those who live chaste lives, and female religious, as Gradus proposes; it is less often recommended for very young women, public women, or married women, for whom it is a better remedy to engage in intercourse with their spouses.

In other words: if a woman got a bit fractious or argumentative, all she needed was genital stimulation, by one means or another.

This attitude to the fairer sex pervaded throughout the centuries, and in Victorian times, many men simply did not take women seriously and did not even consider them as sexual creatures capable of orgasm; much of the medical profession was no better informed.

Following the ancient advice, Victorian doctors advocated genital massage for any fractious woman prone to 'the vapours', a collective term used to describe symptoms that included tiredness, shortness of breath, insomnia, lack of appetite, general irritability or simply disagreeing with one's husband. The archives are full of doctors bemoaning the time spent inducing physician-assisted 'paroxysm', as an orgasm was euphemistically known, because certain uncooperative patients took their time in reaching conclusion.

Extraordinarily, no member of the medical community realized these paroxysms were a female orgasm – apart from the patients, who for the most part proclaimed themselves to feel all the better for their first treatment, immediately agreeing to join the ongoing programme. Consequently, gynaecological massage clinics – they would be called something quite different today – sprang up all over Europe and America.

SUPPLY AND DEMAND

As the demand for such treatments grew to quite alarming proportions, doctors complained of sore fingers and wrists, and fast joined the ranks of the first acknowledged victims of repetitive strain injury (RSI). The Swiss found the solution in a hand-held clockwork device that pummelled adequately well. But the sorry things kept running out of power just as the patient showed signs of impending paroxysm – shortness of breath, flushing of the skin around the neck and the treatment area, and sometimes grunting noises. One can only presume not one single nineteenth-century doctor's wife enjoyed a happy sex life because these professionals failed to recognize the patient's response for what it really was.

Next on the list, and slightly more successful, was hydro-percussion, using water-jets which, when played on the clitoris, brought about rapid and intense paroxysms. Unsurprisingly, it was a hit among patients, and doctors were forced to conclude that hydro-percussion induced a more powerful paroxysm, which was thus all the better for the patient. Demand increased exponentially.

However, the cost of the equipment required and the very fact that a room set up for such treatment was of little use for anything else, put hydro-percussion out of the realm of all but the wealthiest physicians and patients. In stepped the recently revived spa clinics, already in the aqua treatment business, in which doctors set up discreet clinics where ladies flocked to 'take the waters' in their own style – twice daily for one week.

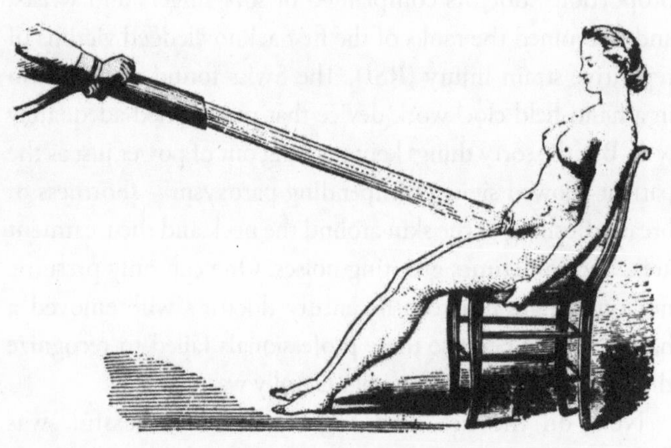

Not for the faint-hearted: hydro-percussion

A SPRING IN HER STEP

The famous French physician Henri Scoutetten's notes from 1843 on hydro-percussion as a remedy for 'female pelvic congestion' revealed why the practice had become so popular:

> The first impression produced by the jet of water is painful, but soon the effect of the percussion, the reaction of the organism to the cold, which causes the skin to flush, and the re-establishment of equilibrium all create so agreeable a sensation that it is necessary to take precautions that they do not go beyond the prescribed time, which is usually four or five minutes. After the douche, the patient dries herself off, refastens her corset, and returns with a brisk step to her room.

MACHINE TAKES OVER

Meanwhile, in 1868 Dr George Taylor of New York, tired of relieving a woman's hysteria by hand – the strain brought on by the number of paroxysms he was required to induce meant he could no longer hold his golf clubs properly – devised a new solution: a treatment table that incorporated a steam-driven vulvar agitator that pounded away from below a durable rubber membrane.

Somewhat cumbersome: the vibrator and the treatment table

It was an immediate hit within the medical profession. Simple to use and with minimal effort involved, all the physician had to do was to instruct his patient to lie face-down on the table with her 'treatment area' positioned over the central aperture. He then fired up the device and asked the patient to make whatever minor adjustments in her position she felt necessary to the successful conclusion of the treatment. But problems with Taylor's trammelling table soon arose – it was large and cumbersome, not to mention noisy, and the recipients of its attentions found it all a bit impersonal; most, apparently, preferred the personal touch.

GRANVILLE'S HAMMER

The solution came in 1880 when Britain's Dr Joseph Mortimer Granville (1833–1900) designed and patented the world's first hand-held electric vibrator for clinical use. He liked to call it his 'percussor' but everyone else, much to the doctor's chagrin, coined it 'Granville's Hammer'.

In profile the percussor resembled a cross between a hair-dryer and an implement a mechanic might use to remove wheelnuts; it could be fitted with a variety of differently shaped rubber heads and was suspended from a moveable gantry when not in use. And best of all, the patients loved it – they flocked to be percussed, again and again. Although Granville himself never used it, stating in his paper 'Nerve Vibration and Excitation as Agents in the Treatment of Functional Disorder and Organic Disease' (1883): 'I have never yet percussed a female patient ... I have avoided, and shall continue to avoid the treatment of women by percussion, simply because I do not wish to be hoodwinked, and help to mislead others, by the vagaries of the hysterical state.'

In 1902 the American market responded to Granville's invention with a less industrial-looking unit for 'self-treatment', sounding the death knell for a majority of the discreet gynaecological massage parlours and for the medical 'pelvic percussing' market. First marketed by the still-extant domestic appliance company Hamilton Beach, the hand-held vibrator was the fifth domestic appliance to be electrified, after the fan, the kettle, the sewing machine and the pop-up toaster.

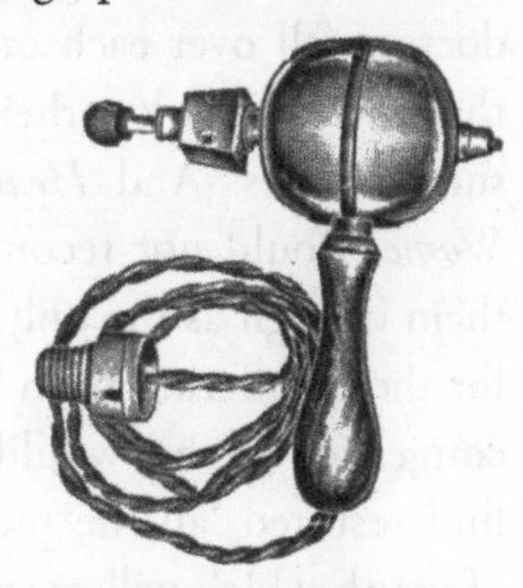

On the move: the portable vibrator

A QUIVERING SUCCESS

And there was gold in them there thrills! Demand was enormous. Widely advertised in respectable publications ranging from the *Sears-Roebuck Catalogue* to the suitably named *Woman's Home Companion*, a whole variety of vibrator models were soon available to suit every budget. These ranged in power from approximately 1,000 pulses a minute on the cheap-and-cheerful models to the top-of-the-range models, including the much eulogized 'Chattanooga', which cost approximately $200 and delivered an eye-watering 8,000 pulses a minute.

The 'Chattanooga' was a free-standing device, roughly 1 metre in height, with a manoeuvrable 'action arm' that could be lowered to the horizontal to render accessible the business end, which was tipped with something resembling a large suppository. Now stripped of their highly lucrative onanistic revenue stream, high-profile doctors fell over each other in the rush to market their own such devices. And *Health for Women* could not recommend them enough as the only resort for those suffering from 'pelvic congestion', who would soon find restored 'all the pleasures of youth which will once again throb within you'.

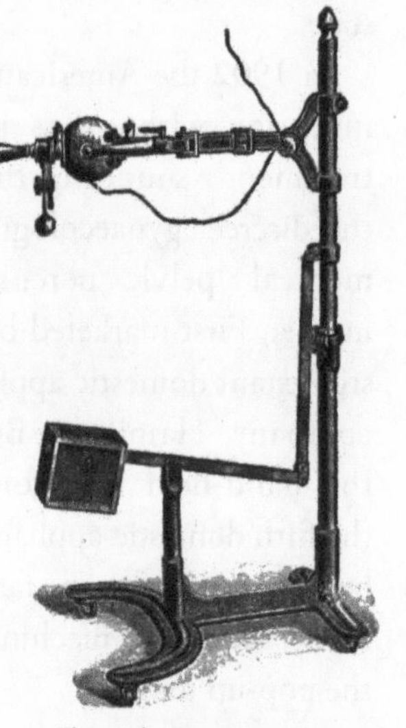

The 'Chattanooga'

NAUGHTY BUT NICE

Thousands of American and European women did indeed throb away to their hearts' content, while the menfolk twirled their waxed moustaches and continued to mutter about 'women's problems'. It may strike the modern reader as incredible that until only the last century thousands of ladies were routinely masturbated by doctors, while their husbands remained oblivious. But then most men, including doctors, were fairly ignorant of female sexuality. Men enjoyed sex and women tolerated it – that was the natural order of things. It is little wonder, therefore, that women nipped out for the odd physician-assisted paroxysm, or invested in a Chattanooga 'choo-choo'.

SERIOUS BUSINESS

All this percussing was not regarded as a furtive and seedy sideline of medicine. The *Merck Manual*, still a respected doctors' guide, listed in its first edition of the twentieth century 'female hysteria' as a recognized condition. 'Pelvic massage', by manual or mechanical means, was recommended as the only effective treatment. And, as proof that no individual, not even at the beginning of the twentieth century, thought there might be a sexual angle to all this percussing, the same manual suggested sulphuric acid should be used to remove sensation from the clitoris of any woman who showed excessive interest in or derived 'too much pleasure' from sexual arousal.

By the 1920s medical masturbation had ceased to be prescribed and ladies' percussers had become hand-held, battery-powered devices, ultimately stripped of their thin veneer of respectability when they made regular appearances in the growing sex-film industry. And finally, in 1952 'hysteria', along with all its attendant symptoms, was removed from all official lists of recognized medical conditions.

All Smoke and Quivers:

Tobacco can cure a variety of health problems

Tobacco takes its name from an early Caribbean word for a cigar (hence the name given to the cigar-shaped island Tobago) and was first brought to Europe from the Americas in or around 1518 by the Spanish – not, as myth would have it, by Sir Walter Raleigh. When tobacco was first introduced to the West it was welcomed as a medicinal miracle that could cure much of that which ails man.

A CURE-ALL

The noxious weed was immediately touted as a wondrous herbal remedy by its importers. They regaled the Spanish and Portuguese courts with tales of the native suppliers eulogizing the benefits of tobacco smoke enemas, in particular. Europeans enthusiastically adopted the practice, and smoke-enema treatments, at the time referred to as 'glysters', became very popular across much of the continent, and would remain so until the mid-nineteenth century.

The case for the many benefits of tobacco was significantly buoyed by the findings of the Spanish physician and botanist Nicolás Monardes (1493–1588). His book on its use in the treatment of a whole host of conditions, ranging from constipation to epilepsy, was published in three parts from 1565 to 1574. As a consequence of Monardes' discoveries, smoke blowing was then used to help cure a range of ills – those afflicted by earache had tobacco smoke blown in their ears; those with sinus problems received smoke up the nose; and those with gastro-intestinal problems by another route

THE FIRST SMOKING BAN

It may come as something of a surprise for anti-tobacco campaigners to learn, but Adolf Hitler was, in fact, the main driving force behind the first-ever national anti-smoking programme. After Nazi doctors discovered the first proof of a link between smoking and lung cancer, and the risk smoking posed to unborn children, Hitler's regime implemented a series of strategies aimed at reducing tobacco consumption.

It was Nazi Germany that banned smoking on public transport, in air-raid shelters and in several public buildings and restaurants, as well as forbidding advertising that presented smoking in a positive light. Smoking was not allowed in the Luftwaffe and SS officers were banned from lighting up while on duty.

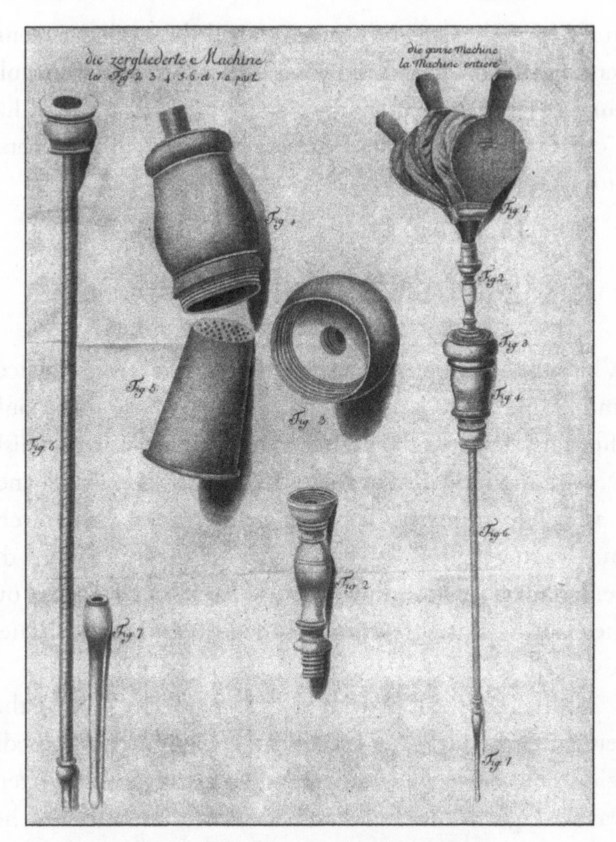

Brace yourself: the tools of the trade

altogether. Although it seems a not insignificant detail was lost in translation: the native tobacco vendors in fact only used tobacco remedies to cure constipation in their horses.

Nevertheless, tobacco smoke remedies became standard practice across Europe; smoke enema treatments the most popular of all. Hard as it may be to contemplate today, from the early sixteenth century until the mid-nineteenth century, the great and the good queued up to have smoke pumped up their rear end by a contraption that looked like a cross between a pair of fireside bellows and an incense burner.

EACH TO HIS OWN

While the common man preferred to take in his tobacco smoke in a more conventional manner, those who could afford it submitted themselves to smoke enemas with relish. Despite the indignity of the procedure, the trend spawned a veritable industry, complete with its own hierarchy and leading lights. At the bottom of the rung were the 'lemonaders', whose unenviable duty it was to clean out the waiting patient with lemonade before the 'fumier' advanced.

An incident in Oxford in December 1650 further cemented the idea that the smoke enema held supposedly miraculous powers. A young servant-girl called Anne Greene was wrongly accused and convicted of the murder of her own child, which had in fact been stillborn. Hanged before the obligatory baying mob, Greene's body was taken down and carted away for dissection by would-be surgeons.

During the procedure someone in the morgue thought they detected a slight twitch of Greene's fingers, so she was immediately administered a smoke enema in an attempt to revive her. It was a success, she sat up in startled confusion and received a full pardon. Greene went on to become a

ALL THAT GLISTERS ...

Despite their surge in popularity, smoke enema treatments were not universally embraced. Shakespeare's 'all that glisters is not gold' from *The Merchant of Venice* is said by some to be a pun based on the practice.

And the treatment did not catch on so well in America. The phrase 'to blow smoke up someone's ass', which refers to an attempted con or deception, was coined in response to the questionable credibility of smoke enemas.

living advertisement for the apparently wondrous powers of the treatment.

While the smoke enema treatment held sway, it was widely respected in the canon of accepted medicine. In 1774 the Society for the Recovery of Persons Apparently Drowned was formed, aimed at promoting lifesaving intervention in the case of drowning. The society used the money received from public subscriptions to instal smoke-enema huts along the London reaches of the Thames and at strategic points on the city's larger lakes. Inspired by Anne Greene, the society wanted to provide people with a failsafe method of either establishing death or of resuscitating victims of near drowning. The society did in fact enjoy numerous successes and, from such unlikely beginnings, it evolved into the present-day Royal Humane Society.

DEARLY DEPARTED

The notoriety of Greene's enema also gave rise to the habit within moneyed circles to pump the recently departed with smoke, 'just to make sure' – even in death there was no escaping the smoking glyster. In the early nineteenth century, when the practice had fallen from grace, those fearful of being buried alive could have a bell-string installed in the coffin to alert graveyard attendants to their predicament. This practice did not, as some maintain, give rise to expressions such as 'dead-ringer' and 'saved by the bell', nor 'graveyard shift' – as worked by those who kept nightly vigil, listening out for any lonely tinkling.

PURE POISON

Anal fumigation went into decline in the early nineteenth century as scientific research began to reveal tobacco's poisonous qualities. The English physiologist and surgeon Sir Benjamin Brodie led the most significant research in this area, when he discovered that nicotine, the principal ingredient in tobacco, can interfere with blood circulation. Nevertheless, mainstream medical thought still considered tobacco smoke capable of eradicating cholera. No matter how improbable this sounds, it must be remembered that these were the days before the broad acceptance of

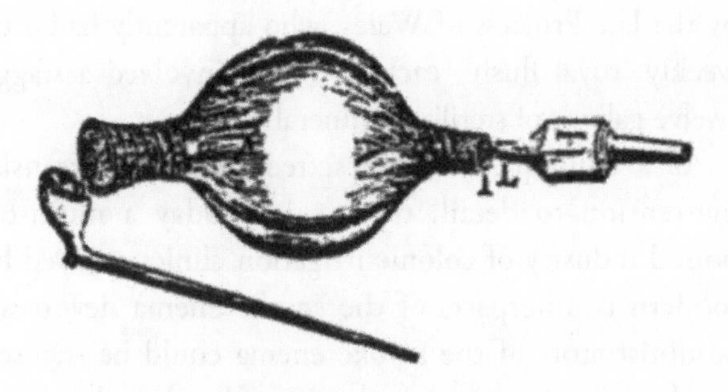

A device for administering smoke enemas

germ theory, at a time when all disease was thought to be transmitted by foul smells (as in malaria – which literally translates as 'bad air', see Heaven Scent on page 106). Free tobacco was issued following every outbreak of cholera, with all recipients, including children, required to puff away to create a fug to fight off the cholera 'fumes'.

COLONICS CATCH ON

Despite the treatment's general decline by the mid-nineteenth century, smoke enemas had a lasting legacy – as the fumiers disappeared, the lowly lemonaders rose to take their place. People had grown used to, and even obsessed with, things being shoved up their bottoms, and they seem to have decided that even if the smoke had to go, why not stick with the lemonade wash. Although there is evidence to suggest the Ancient Egyptians and Greeks favoured such bizarre 'hygiene', this was the beginning of the still-continuing and rather questionable trend for colonic irrigation – as championed by the late Princess of Wales, who apparently had a thrice-weekly 'royal flush', each of which involved a staggering twelve gallons of sterilized mineral water.

As a consequence of a sixteenth-century translator's inattention to detail, there exists today a multi-billion pound industry of colonic irrigation clinics enjoyed by the modern counterparts of the smoke enema devotees. The administrators of the smoke enema could be excused for not knowing any better; they actually thought they were administering a valid treatment. The same, however, cannot be said for the modern-day irrigation lobby, which has managed to convince its patients that we are all being slowly poisoned by impacted faecal matter sticking to the inside of the large intestine.

No autopsy performed since records began has ever found evidence to support this, and the pointless and, it could be argued, dangerous practice of irrigation can cause side effects ranging from amoebic infection to internal perforations and heart failure. Personally, I find that six pints of strong ale administered from the other end of the body is a safer and far more enjoyable option.

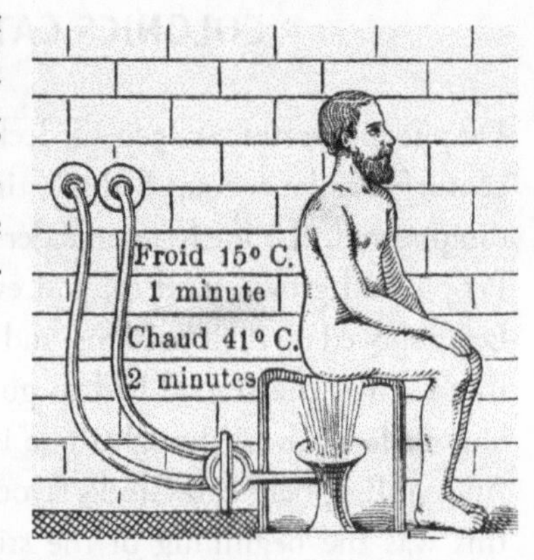

In the wake of the smoke

Hey, Hey, We're the Monkeys!

Injecting monkeys' glands into humans encourages sexual rejuvenation

FOR CENTURIES PEOPLE, or, rather, men, have been searching for a substance that will either render women so sexually aroused that they will succumb to the advances of any man, or one that will reawaken the most jaded of loins. Until the invention of Viagra (although technically not an aphrodisiac) in the late twentieth century, the most famous sexual stimulant was Spanish Fly. Prepared from the wing-sheaths of the blister beetle, the aphrodisiac has been used and abused by the jaded and lascivious since Roman times.

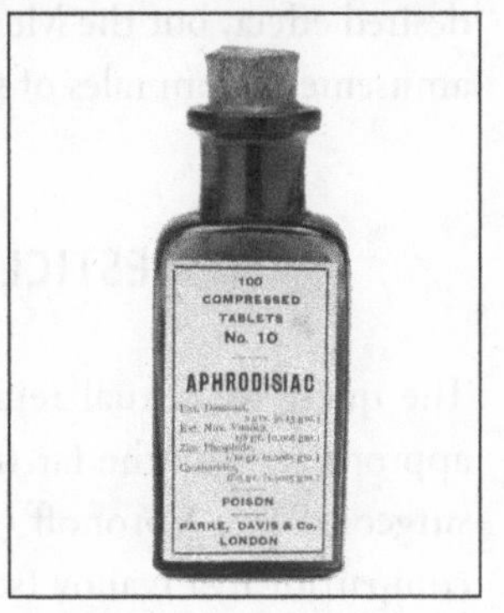

A helping hand: an aphrodisiac containing Spanish Fly

How its reputation has endured is, however, a mystery as the closest to any aphrodisiac effect it engenders is irritation of the urinary tract, with other side effects ranging from vomiting, diarrhoea and permanent damage to the kidneys to cardiac arrhythmia and death. Far from lighting people's fires, Spanish Fly would have put the dampers on any Roman orgy; and yet it is still touted in liquid and tablet form on the Internet today for those who like to live dangerously – or not at all.

FOOD OF LOVE

Over the centuries, various foods have enjoyed transient notoriety for their supposed aphrodisiac effects. Sixteenth-century Spanish explorers were entranced by the pendulous avocado fruits of Mexico, especially when they were told 'avocado' meant 'testicle' in the local tongue. The amorous conquistadores immediately began shipping the fruits home, where they would be turned into a paste for foolish old men to apply to their genitals while sitting out in the sun. Naturally the crushed avocados failed to have the desired effect, but the Mexicans, it seems, derived no end of amusement from tales of such antics.

TESTICLE TREATMENT

The quest for sexual rejuvenation continued; soon to be appropriated by the far-from-ethical Russian-born French surgeon Serge Voronoff (1866–1951) who, along with his compatriot Ilya Ivanov (see Origin of the Specious on page 125) may have been responsible for the deaths of millions of people worldwide.

THE HUMBLE TOMATO

The tomato once enjoyed extended infamy as a potent aphrodisiac. The fruit was introduced to Europe by the Moors, which prompted the French to call it *pomme de maure*. The English misheard this as *pomme d'amour*, or love apple, and leapt to the inevitable conclusion. The medieval Church mounted a counterattack by suggesting the fruit was in fact poisonous.

Because the tomato plant resembles a deadly nightshade (to which it is in fact related), notions of the fruit's so-called toxicity became enshrined in medical lore. The idea became cemented in the sixteenth-century mind with the publication of John Gerard's *Herball* (1597), which condemned the plant for generations to come. Long after the Church had moved on to other concerns, the medical profession was still firmly of the opinion that anyone who ate two or three tomatoes would suffer instant death. It was not until the early-to-mid-eighteenth century that this idea was on the whole disproved following simple experimentation on the part of thrill-seekers.

Americans remained convinced of the tomato's toxicity until the early nineteenth century, when, according to legend, Colonel Robert Gibbon Johnson stood on the steps of the Old Courthouse of Salem on 26 September 1820 and ate a basketful before an astonished crowd.

In 1889, while conducting pioneering research into methods of curbing the ageing process, Voronoff began to practise his methods on himself, injecting his body with extracts from ground-up dog and guinea-pig testicles. The effects were barely perceptible, so Voronoff moved on to glandular tissue transplants of the same substance, and published papers eulogizing it as the way forward in the fight against everything from flagging desire to schizophrenia. The international press and the European and American medical profession naively accepted Voronoff's findings without ever asking him for corroborative evidence.

In early 1920 Voronoff began transplanting the testes of executed criminals into the gullible who had cash to spare. When supply failed to keep up with demand, he began using monkey testicles, grafting thin slices of the glands into the scrotums of the rich and famous. By 1922, monkey gland treatment was the talk of the medical profession. Voronoff grew rich, not only from the operations he conducted on the likes of Kemal Ataturk, the first President of the Republic of Turkey, and other heads of state, but also from the income he generated by teaching other doctors and surgeons to expand his folly across the rest of Europe and America.

In 1923 more than 700 high-status delegates of the International Congress of Surgeons came from all over Europe and America to laud Voronoff at a conference in London and to hail him as the father of rejuvenation. However, no person in attendance seemed to notice that the master himself looked a trifle aged and balding; or if they did, no one had the courage to cry, 'Physician, heal thyself!'

IT FELT LIKE A GOOD IDEA AT THE TIME

In the late 1780s bunnies were introduced to Australia. The fox was introduced 100 years later to deal with the rabbits' expanding population – but instead of eating the rabbits, the foxes preferred to eat the sheep. Eventually, in the 1950s the vile myxomatosis was unleashed to undo what should never have been done in the first place.

DISAPPOINTING RESULTS

By 1930 Voronoff had packed monkey genitals into the scrotums of over 500 wealthy patients in France alone, but he would soon overplay his hand. Diversifying into the female market, Voronoff began transplanting monkey ovaries into women who feared the onset of age. At first, women flocked to his surgery door as the carrot of lasting allure proved too strong to resist. The results were disappointing: none of Voronoff's female patients saw any discernible retardation in their natural ageing. To make matters worse, many of the men Voronoff had treated in the early 1920s began to die in large numbers; dark mutterings of disillusionment and dissent grew louder.

It then became public that Voronoff had transplanted a human ovary into a female monkey and had inseminated it with human sperm. This was a step too far for his now less-than-adoring public. The hitherto sycophantic medical profession also found itself getting twitchy, and it started to shine the harsh light of scepticism, which should have been directed long before, on to Voronoff's work and claims.

By the close of the 1930s, not only were most of Voronoff's early patients six feet under – none having fathered broods of children late in life, nor having lived past any conventional age – but testosterone had been synthesized, enabling direct-injection comparison tests to be conducted. As these tests were progressing, Voronoff, knowing full well what the results would be, quietly prepared for a life of opulent retirement in Switzerland.

As Voronoff expected, none of the experiments he had claimed to have conducted on the rejuvenation of farmyard animals could be replicated. The whole charade might have been quite amusing were it not for one troublesome detail. The modern scourge of HIV is not a disease of the 1980s (see Origin of the Specious on page 126) but one that first arose in the late 1920s when SIV, the simian equivalent, somehow jumped the species barrier. Not surprisingly, there are many today who work and research in that field of medicine who consider the transplanting of monkey reproductive organs into assorted humans to have been a possible vector.

From Mendel to Mengele

The selective breeding of humans can weed out the weak from society

CHARLES DARWIN (1809–82) could never have foreseen the long-term ramifications of his published works. In the short term the fall-out was bad enough. Pilloried by members of the Church who, along with other ill-informed antagonists, had not even read his work properly, Darwin was castigated for proclaiming man is the descendant of monkeys. In reality, he had not proposed anything of the sort. In the long term, the effects were far more devastating. 'Survival of the fittest', a phrase attributed to Darwin, was later used by tyrannical elements in justification of, among other oppressive policies, the new 'science' of eugenics. Darwin had in fact never used the expression; and the man who *had* coined the phrase, the English biologist, philosopher and sociologist Herbert Spencer (1820–1903), had in fact intended it to mean those creatures best fitted to their environment, be they weak *or* strong.

EUGENICS IS BORN

The most questionable scientific idea based on Darwin's work arose when his own cousin, Francis Galton (1822–1911) – the man who championed the use of fingerprints in criminal detection – used Darwin's work as the basis for eugenics. Derived from the Greek *eugenes* (of noble race or birth), eugenics advocated controlled breeding in an attempt to increase the chances of desirable characteristics in offspring.

Like many intellectuals, Darwin spoke before considering the repercussions. In his *The Descent of Man, and Selection in Relation to Sex* (1882) he mused on how medical and scientific advances had meant that the weaker and less productive of our species were artificially propped up to allow them to survive and breed; a harsher environment would naturally cull such parasites. At his most incendiary, Darwin suggested:

> Thus the weaker members of civilized societies propagate their kind … No one who has attended to the breeding of domestic animals will doubt that this must be highly injurious to the race of man … but there appears to be at least one check in steady action, namely that the weaker and inferior members of society do not marry so freely as the sound; and this check might be indefinitely increased by the weak in body or mind refraining from marriage, though this is more to be hoped for than expected.

Within a matter of months of reading his cousin's book Galton had formulated his own take on the future

Eugenics makes the world go round: the front cover of America's satirical *Puck* magazine, June 1913

of humanity. British society – and indeed the entire world – would benefit enormously if all such dead wood was eliminated. His *Inquiries into Human Faculty and its Development* (1883) first coined the term eugenics.

GENETICS HOLDS THE KEY

It all made perfect sense: breeding kennels always match the strongest and the smartest dog with the smartest and the best bitch, and the equine bloodstock lines had been running on the same principles for centuries. New discoveries in the science of genetics further helped Galton's case. Gregor Mendel's experiments on pea patches (see box opposite) led to an understanding of how heredity works. When Mendel crossed pea plants with defined but opposite characteristics – for example a long and a short stem – the result was not an average of the two heights, but a tall plant. In an outright challenge to contemporary scientific thought that believed offspring inherited a blend of their parents' characteristics, Mendel's discoveries concluded that inherited characteristics can be passed on to offspring unaltered, with the strongest predominating. Galton's eugenics applied Mendel's findings to the human race: why not selectively breed together the very best and in the process weed out the worst of the species from the gene pool?

While Galton did not suggest that extant individuals deemed to be defective should be eliminated, he did suggest they be sterilized in order to prevent them from propagating any more of their kind. Nobility of spirit, intelligence and

THE FATHER OF GENETICS

The Abbot of the Augustine Abbey of St Thomas at Brno, Gregor Mendel (1822–84), conducted research into peas grown in the abbey grounds, and his findings awarded him the posthumous recognition as the founder of genetics. Mendel conducted his experiments between 1856 and 1863 and the results led him to devise his two laws of inheritance.

The first law (The Law of Segregation) stated that an individual possesses two alleles (a different form of a gene) for any given trait, one of which is passed on from the mother, the second from the father; whichever of these two alleles is dominant determines the character of the offspring. The second law (The Law of Independent Assortment) stated that separate genes for separate traits are passed on independently of each other. When Mendel published his results in 1866 they were met with derision. It was not until the turn of the century that the 'Father of Genetics' was rediscovered.

artistic talent were, he decided, all inherited traits, as indeed were fecklessness, imbecility, promiscuity, drunkenness and criminality; it would be no different to the selective breeding of dogs or bloodstock. Galton promised that within the span of a few generations crime and anti-social behaviour would be a thing of the past and Britain would be left teeming with pleasant people who bore increasingly talented offspring.

GALTON'S SUPPORTERS

Many of the great and good of Europe and America flocked eagerly to follow Galton's banner. Notable figures including Winston Churchill and Theodore Roosevelt were open and ardent supporters of Galton's movement, as were birth-control activists Marie Stopes and Margaret Sanger. Economists such as John Maynard Keynes and Sidney Webb, the founder of the London School of Economics, saw the financial sense of a society unfettered by the financial burden of supporting an ever-increasing number of unproductive dependants. The American proponent of moral and edible fibre, John Harvey Kellogg, was also in support of anything that improved the purity of the species. Indeed, his most famous product was initially intended to be an anti-masturbation measure – a plain diet, so Kellogg thought, would dampen passionate thoughts.

LEFT-WING SUPPORT

Many today dismiss the eugenics bandwagon as an exclusively right-wing vehicle, but this was far from the case. Virtually every member of the Fabian Society, from which emerged the Labour Party, was an ardent and vocal supporter, including Irish poet and playwright W.B. Yeats (1865–1939), the leader of the suffragette movement Emmeline Pankhurst (1858–1928), Labour Prime Minister Ramsay MacDonald (1866–1937) and economist and social reformer William Beveridge (1879–1963).

Irish playwright and co-founder of the London School of Economics George Bernard Shaw (1856–1950) was convinced that the future of Socialism lay in what he called Social Darwinism and the 'selective breeding of man'. The philosopher Bertrand Russell (1872–1970) went one step further when he proposed that the state should issue

IN EVERYBODY'S WELFARE

The image that most hold of William Beveridge is that of a kindly man whose Report in 1942 gave rise to the British Welfare State. However, the social and health institutions we have today bear little resemblance to the selective ones that Beveridge had in mind. While he appreciated that the state should support those unable to find work, Beveridge believed those who received benefits should lose 'all citizen rights – including not only the franchise but civil freedom and fatherhood'.

Beveridge's plan was for the whole state support structure to be crafted in such a way as to encourage the breeding of the middle and upper classes, who would receive far more benefits than the lower orders, who would be repressed within their own strain. On the very night his Report was being debated in Westminster, Beveridge addressed the Eugenics Society to assure the nervous members that this should indeed be the result. Mercifully this was not how the Report eventually went through in 1945.

everyone with colour-coded 'procreation tickets'; anyone caught having sexual relations with a partner holding a differently coloured card should be given a hefty fine or even face imprisonment for 'genetic treason'.

CALIFORNIA DREAMING

Meanwhile, in America eugenics had gathered significant momentum, and would go on to influence Hitler, the man who took it to its sickeningly logical conclusion. The concept of a blond-haired, blue-eyed Nordic super-race did not originate with the Führer; he derived the idea from studying the Californian eugenics programme that kicked off in 1909. It was the first state to enshrine eugenics principles in its legislation, and the Californian eugenics programme allowed for the enforced isolation and sterilization of unfit individuals (with the definition of 'unfit' left wide open to interpretation), and marriage restriction laws. Before the eugenics programme was finally eradicated in America, over 60,000 'unfit individuals' would be forcibly sterilized and as many marriages ruled illegal, with California alone accounting for roughly one third of that total.

Without the financial muscle of the Carnegie Institution, the Rockefeller Foundation and that of countless industrial magnates, the Eugenics Programme of America would doubtless have floundered. It also received verbal backing from a majority of the Ivy League institutions, with Stanford's President David Starr Jordan's *Blood of a Nation* published in 1902 in support of the eugenics movement. In

1904, the Carnegie Institution began funding the Eugenics Records Office (ERO) at a laboratory complex on Long Island. It was there that millions of index cards catalogued the lineage and identified patterns of inherited conditions of American citizens, which the ERO used to justify its, often successful, demands for the expansion of eugenics legislation and an intensification and broadening of the sterilization schemes. The computer technology company IBM would later copy these record-keeping methods when it developed a punch-card system to help Hitler run his own eugenics programme. The infamous tattoo on concentration camp inmates' inner forearms was not just a numeric ID, it was their IBM number in which was coded race, deviancy and skill – 'Dutch, communist, carpenter', for example.

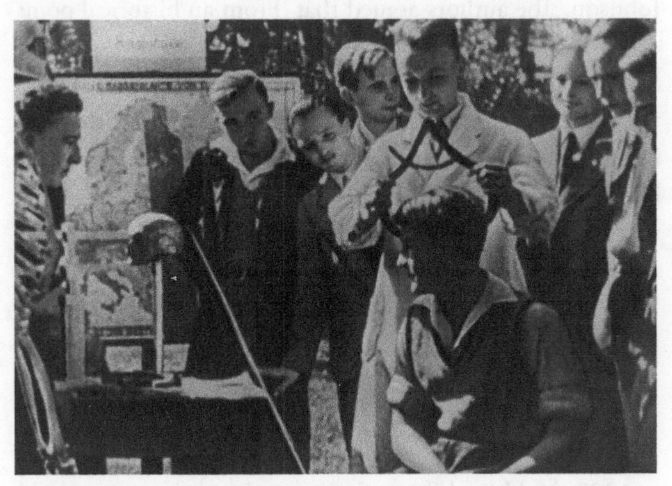

This still from the 1965 film *Ordinary Fascism* shows a man's head being measured to test for Aryan qualities

EUGENICIDE

American eugenics took a seriously dark turn with the publication in 1911 of the Carnegie-supported preliminary report by the American Breeders' Association on 'The Best Practical Means for Cutting off the Defective Germ-Plasm in the Human Population'. This presented an eighteen-point agendum in which point eight explored the use of euthanasia for the most hopeless of all cases. The commonly suggested form of 'eugenicide' was by lethal gas chamber, a term and 'solution' that would soon become all too distressingly familiar.

In 1918 Paul Popenoe (1888–1979), prominent eugenicist and US Army medical specialist in venereal diseases, co-authored *Applied Eugenics* with Roswell H. Johnson. The authors argued that 'From an historical point of view, the first method which presents itself is execution and its value in keeping up the standard of the race should not be underestimated.' Popenoe was convinced that the infanticide employed in Ancient Rome and Sparta, as a birth-control method for unburdening the state of infants that showed signs of weakness or physical imperfection, was a model well worth exploring.

BUCK v BELL

The 'science' of eugenics showed little sign of waning when in 1928 the United States Supreme Court chose to uphold a statute that promoted the compulsory sterilization of the unfit during the landmark case of *Buck v Bell*. The case centred on rape victim Carrie Buck (1906–1983) who had been committed after the attack to the Virginia State Colony for Epileptics and the Feeble-Minded by her foster parents, John and Alice Dobbs. With their own nephew the prime suspect, the Dobbs family raced to have Carrie committed as 'incompetent and promiscuous', with the eugenics lobby hotly demanding her sterilization. They alleged her natural mother had also been promiscuous and 'sub-normal' and, although untrue, the court accepted the slur without question.

The case sat before Justice Oliver Wendell Holmes, the acclaimed physician, writer and poet. He sanctioned the enforced sterilization, pronouncing, 'It is better for all the world, if instead of waiting to execute degenerate offspring for crime, or to let them starve for their imbecility, society can prevent those who are manifestly unfit from continuing their kind. Three generations of imbeciles are enough.'

This was both unfair and untrue: the child born as a result of the rape, Vivian, was a perfectly normal girl who did well at school but died, aged eight, of enterocolitis. Nevertheless, close tabs were kept on Carrie's other relatives, most probably by the ERO. When Carrie's sister, Doris, was admitted to hospital for an appendectomy, the surgeons were alerted to her 'red-flag' status and automatically sterilized her while she was under anaesthetic in order to 'terminate the family pollution of society'.

Tragically, Doris was not even told of the procedure on waking; later, medical consultation occasioned by her not conceiving when married revealed all. Carrie lived into her seventies, a life-long and avid reader, and those who found themselves partnered against her at bridge had no call to think her 'feeble-minded'. Justice Holmes' shameful summing-up of the case was later quoted back to the American judges presiding over the Nazi Nuremberg war trials by the defendants' counsel.

A MATCH MADE IN HEAVEN

In 1934 the founder of the California State University at Sacramento and leading light of the Californian eugenics programme, Charles M. Goethe (1875–1966), accepted an invitation from Germany to observe its developments in the area of eugenics. Although Germany's population was far smaller than that of the United States, its sterilizations were already by then exceeding 5,000 per month. Upon his return to California, Goethe gathered together his fellow eugenics committee members to congratulate them:

> You will be interested to know that your work has played a powerful part in shaping the opinions of the group of intellectuals who are behind Hitler in this epoch-making program. Everywhere I sensed that their opinions have been tremendously stimulated by American thought. I want you, my dear friends, to carry this thought with you for the rest of your lives, that you have really jolted into action a great government of 60 million people.

They must have been so proud.

But the darkest link between the established American eugenics programmes and Germany had in fact been forged earlier, when the Rockefeller Foundation had helped to establish the German eugenics programme. The foundation had sent approximately $4 million (at today's value) in donations to several dubious German research projects; the Kaiser Wilhelm Institute for Anthropology, Human Heredity and Eugenics in Berlin was the major beneficiary.

For quite some time the American eugenics lobby had been thwarted in its dubious desires to conduct experiments on twins, but with Hitler they saw the chance to do so by proxy. On 13 May 1932 the foundation's New York office cabled its Paris counterpart with the following missive: 'JUNE MEETING EXECUTIVE COMMITTEE NINE THOUSAND DOLLARS OVER THREE YEAR PERIOD TO KWG INSTITUTE ANTHROPOLOGY FOR RESEARCH ON TWINS AND EFFECTS ON LATER GENERATIONS OF SUBSTANCES TOXIC FOR GERM PLASM.'

At the time, the Head of the Institute of Anthropology, Human Heredity and Eugenics was Otmar Freiherr von Verschuer (1896–1969), a man well known in American eugenics circles, and whose gruesome assistants Josef Mengele (1911–79) and Karin Magnussen (1908–97) would later achieve their own notoriety. But it was here, under Verschuer and with Rockefeller money, that Mengele started on the road that led him to the SS and Auschwitz, and further unspeakable experiments on twins.

As the Second World War drew closer, the Rockefeller

Foundation suspended all funding, but the Verschuer-Mengele programme had by then assumed a life and will of its own. From Auschwitz, Mengele would send off all manner of tissue and blood samples from twins he had infected with diseases ranging from typhus to syphilis. To Magnussen he sent twins' eyes, especially those of any pair unlucky enough to have eyes of a different colour to one another, her particular area of interest. Extraordinarily, and despite screeds of evidence pointing to Verschuer and Magnussen's culpability, they were both spared prosecution as war criminals.

WELL I NEVER! POPULAR SCIENTIFIC IDEAS DEBUNKED

- When your arm goes to sleep it is not through inhibited blood flow.
- Eyeballs cannot be removed from their sockets for treatment and then popped back in again.
- Sugar does not make kids hyperactive.

In July 1946 when the war was over, Popenoe and Verschuer resumed their correspondence, with the former informing the latter, 'It was indeed a pleasure to hear from you again. I have been very anxious about my colleagues in Germany. I suppose sterilization has been discontinued [there]?' Some people just don't know when to give up.

Within a matter of a few years, and like so many other American and German eugenicists, Verschuer had successfully rebranded himself as a geneticist. He assumed a comfortable professorship at the University of Münster, became a valued member of the American Society of Human Genetics, and yet all the while he maintained, until his death, that he had accepted his membership to the American Eugenics Society *during* the war.

THE QUEST CONTINUES

If there is one thing that man learns from experience is that man does *not* learn from experience. Eugenics is dead; long live 'newgenics'. Recent genetic developments see us yet again standing at the threshold of human perfection. 'De-selection' – what a wonderfully innocuous-sounding term. Yet nearly 2,300 abortions of foetuses with mental and physical disabilities were carried out in the UK alone in 2010.

It is all too easy to be seduced down this route. Who will decide who lives and who dies, and who will set the parameters? All who think they are fit for making such decisions should first be made to watch all available footage of Mengele and his like as they dealt with and disposed of those they considered to be genetic trash. Their abhorrent crimes were not committed centuries ago; nor on another planet – Mengele's experiments were carried out a scant sixty-five years ago on European soil.

Plane Stupidity

The earth is flat

THE THEORY OF a flat earth immediately brings to mind the detractors who had derided Christopher Columbus' momentous 1492 voyage to claim a selection of Caribbean islands for Spain, when they suggested he would simply sail off the edge of the world.

In reality, very few people in Columbus' time thought the earth was flat; the notion that it had been an accepted thought was invented in the mid-nineteenth century by the American humorist Washington Irving (1783–1859) in his hugely popular book *The Life and Voyages of Christopher Columbus* (1828). Irving concocted an entirely false account of Columbus' confrontation with the Salamancan Committee, complete with fabricated quotes attributed to idiotic clerics who allegedly spouted spurious notions of a flat earth.

As events transpired, the Salamancan Council objected to Columbus' gross underestimation of the expanse of water he was proposing to cross. They were right and Columbus wrong – the world was twice the size of the explorer's estimations.

Columbus – with globe – before the court

Nevertheless, the notion of a flat earth was both ancient and all pervasive, and it is also still adhered to by some. Perhaps the inspiration for Irving's prank was the fact that the medieval Church was chock-full of hidebound flat-earthers.

ELEPHANTS AND TURTLES

Hindu cosmology suggested the earth was a flat-bottomed dome that was carried on the back of four elephants, which in turn stood on a giant turtle that swam in an infinite ocean. The Babylonians too were flat-earthers. They believed the globe was a disc floating in the sea, surrounded by a rim of mountains that supported the heavens. The ancient Egyptians also believed the earth to be flat, except their version saw the earth as rectangular shaped, with Egypt, naturally enough, situated in the centre.

In much the same way, during the Middle Ages early European flat-earthers believed in a flat and square earth because the Bible (Rev. 7:1) makes reference to the four corners of the earth, each of which was guarded by angels who controlled the four winds. This was a time when the Church reserved the right to incinerate alive any who challenged the very word of the Bible. Although the sane of mind realized a sphere with corners is an impossiblity, those who feared for their lives wisely nodded and paid lip-service to the concept of a flat earth.

The world according to Hindu cosmology

AGAINST THE ODDS

It is difficult, with hindsight, to see how such folly was allowed to triumph over proof to the contrary. The Ancient Greeks had realized the earth was a sphere simply by observing how inbound ships were first visible by the tip of their masts, with the rest of the vessel rising into view as it drew closer. They were extensive travellers and soon deduced that as their location changed, so too did their relationship to the stars, with some disappearing from view while new ones appeared. They also observed that no matter how far they sailed, the inclination all around them, from the horizons to the skies, remained the same. This cone of inclination they called the *klima*, hence 'climb', and the changing weather within any one such cone was the climate. Come the Middle Ages, few with any intelligence thought the Greeks wrong, but few with a love of life were prepared to defy the Church.

Pope Gregory the Great (540–604) openly proclaimed it to be heresy to denounce the concept of a flat earth, and as late as the sixteenth century Pope Alexander VI (1431–1503) was still maintaining the same stance. Better known as Roderic Borja, Alexander VI had no difficulty reconciling his penchant for rape and murder with his Christian ethics but, although a somewhat savvy and hand-to-throat political survivor, Borja lacked intelligence.

He was also pretty greedy and by 1493 he had wearied of the two leading Catholic nations, Spain and Portugal, fighting each other over new territories during the Age of Discovery – while the countries fought between themselves they were failing to fill Borja's Vatican coffers. As a dedicated

JUDGEMENT DAY

Flat-earthers were not all on the Catholic side of the fence. Although the German monk Martin Luther (1483–1546) today enjoys a reputation as a gentle, pious and measured man, he was in fact a deeply unpleasant and small-minded anti-Semite who thought that all Jews should be covered in excrement and whipped out of town; he thought too that the poor and downtrodden should be treated no better because that is what God intended. Of extremely blinkered perspective, Luther was also a flat-earther. His reasoning stemmed from the fact that all humanity was destined to witness The Second Coming on Judgement Day and, if the earth was a sphere, the view of the majority of the planet's population would be obscured.

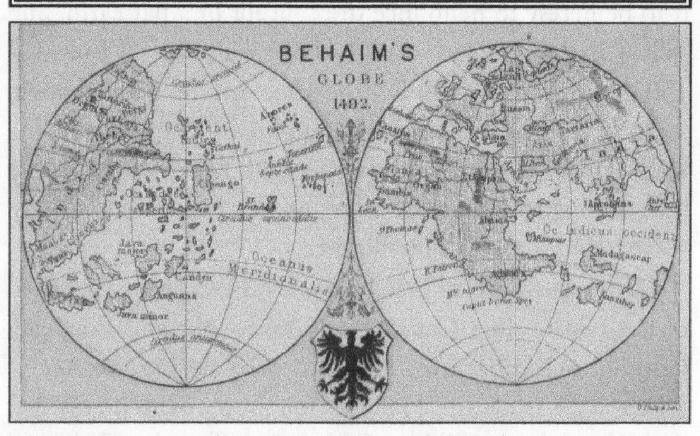

This illustration of a globe, dated 1492, rejects the theory of a flat earth

flat-earther, Borja decided to draw a line down the map (the first to be called a Demarcation Line) and pronounced that the Portuguese hunt to the east of the line and the Spanish to the west. (The line cut through South America, which is why Portuguese is spoken in Brazil, and Spanish elsewhere.) But it never occurred to Borja that the Portuguese could simply keep sailing east to end up in the west.

DETRACTORS IN THE MIDST

One of the few who dared to speak out against the flat-earth theory was the Portuguese explorer Ferdinand Magellan (1480–1521) who, prior to embarking on his famous circumnavigation of 1519, pronounced, 'The Church says the earth is flat but I have seen its shadow on the Moon and I have more faith in that shadow than I do in the Church.' Having spoken so boldly, Magellan had the presence of mind to leap smartly aboard his ship and set sail. Eventually, as circumnavigations became more frequent, the Vatican was forced to abandon its stance and admit that perhaps the earth was a sphere. Flat-earth thinking staggered on in a few fringe Fundamentalist groups, mainly in America, but strange concepts of the form and profile of the earth were not dead – even in the twentieth century.

JUMPING ON THE BANDWAGON

Far from being a rational man, Hitler too explored the flat-earth theory (as well as, quite conversely, the hollow earth

An artistic representation of Columbus' arrival at Margarita Island in 1498

theory, see Subterranean Homesick Blues on page 155) and finally decided that, although not flat, the earth was perhaps concave. He believed humanity ran around, upside down, on the interior of the surface, much like an army of ants. To explore this notion he ordered the German-born scientist Wernher von Braun (1912–77) to fire off a series of rockets at a 45-degree angle in order to see if they landed in Australia. As the Church of old had been before him, Hitler was by this point very much in charge, so von Braun nodded solemnly and attempted the doomed mission. Naturally nervous of reporting his failure to hit down under, von Braun gently guided the Führer to the notion that, while the theory of a flat earth may well be correct, the rockets did not contain the power necessary to validate it. Hitler seemed appeased and von Braun made a sharp exit. But the Führer's fascination did not end there.

In April 1942 Hitler organized another attempt to prove himself right. Under the leadership of Dr Heinz Fischer, a specialist in the field of infrared rays, an expedition was dispatched to Rügen Island in the Baltic Sea to attempt the same experiment, but this time with the aid of radar. For weeks on end the radar scanned at a 45-degree angle but nothing came back, leaving Fischer and his team extremely nervous about returning home. But they did not have to worry; by this time Hitler's attentions had been diverted to the invasion of Russia, which allowed the members of the expedition to slip quietly back into Germany unnoticed.

THE FINAL FRONTIER: SPACE GETS SILLY

Space was once home to the following daft theories:

- Canals exist on Mars.
- The moon is a hollow vessel.
- The universe is static and does not expand.
- The planet Vulcan orbits between Mercury and the sun.
- A full moon causes lunacy.

From Popcorn to Mozart

Humans can be influenced by subliminal messaging

OF ALL THE discredited theories described in this book, the one explored in this chapter is not only the most modern, but also responsible for encouraging more spin-offs of spurious science than any other.

MAD MEN

America during the late 1950s was a consumer's Temple of Mammon in which marketing men served as high priests. In 1957 market researcher James Vicary (1915–77) arrived on the scene, holding aloft the invisible Holy Grail of advertising.

Vicary reported the incredible successes he claimed he had achieved in conducting covert experiments in a public cinema in Fort Lee, New Jersey in the summer of that same year. Vicary quite logically reasoned that consumers possessed a personal level of in-built sales resistance. He suggested that while their heart may want the product, the ad-man's greatest enemy – the consumer's brain – was sitting in ultimate control, warning the owner. So, argued Vicary, why not cut the conscious brain out of the equation?

Vicary claimed that over a period of sixteen weeks he had exposed 45,000 cinema-goers to doctored versions of whichever film they had gone to watch. Using a tachistoscope, an instrument used to display an image for a brief period, Vicary had incorporated flash-images that instructed audiences to 'drink Coke' and 'eat popcorn'. Supporting his research with genuine science as to how fast the human eye can capture images and why, Vicary informed Madison Avenue ad gurus that his images endured for 1/3000 of a second – too fast for the eye and the conscious brain to notice, but detectable nonetheless on a subliminal level.

Hungry? Eat popcorn: a mock-up of Vicary's experiment

Because the brain had been influenced on an almost hypnotic level, no defensive actions had been activated. The result, claimed Vicary, had been an increase in the foyer sales of Coke and popcorn: they had risen by 18 per cent and 57 per cent respectively. American manufacturers were agog; it all seemed to make perfect sense. The commercial worth of the carrot dangled before their collective eyes was deemed to be inestimable.

LIGHTS, CAMERA, ACTION: PHYSICS-GONE-WRONG IN THE MOVIES

- Victims of gunfire shots would never be flung backwards because every action has an equal and opposite reaction.
- There is no such thing as a silencer that can reduce the noise of a gunshot to a muted 'phut'.
- Quicksand does not cause drowning, because it's hundreds of times more buoyant than the Dead Sea.
- People ejected into space don't explode or experience boiling of the blood.

POLITICAL TWIST

Others, too, were excited by the possibilities offered by Vicary's discoveries, although their agendum hid an altogether more sinister edge. Political parties debated the possibility of using such tactics not only in their own television broadcasts, but also during popular shows that were unrelated to politics.

There was a sudden and undignified scramble in the political arena as vast salaries were offered as incentives to entice opinion-makers and so-called Depth Men (psychological manipulators) away from their Madison Avenue desks and up to Capitol Hill. Journalist Vance Packard published his still-famous *The Hidden Persuaders* (1957), which linked Washington's use of commercial advertising tactics to Vicary's experiment. Science-fiction author Aldous Huxley told anyone who would listen that his *Brave New World* (1931) had come to hideous life, with Vicary 'sounding the death knell of free will'. And, as the debate over the ethicality of subliminal messaging rose to fever pitch, Richard Condon published *The Manchurian Candidate* (1959). A thriller that features the brainwashing of individuals by political parties, Condon's novel fanned the flames of conspiracy.

BRAINWASHING

Vicary could not have picked a better time for his fraud – for fraud is what it was – as America was still smarting from recent revelations that had emerged after the Korean War (1950–53). The sensationalist writer Edward Hunter's *Brainwashing in Red China* (1951) – which introduced the term 'brainwashing' to the American people – induced paranoia amongst Americans by presenting images of shadowy Fu Manchu-like characters manipulating the minds of American PoWs.

In reality, there had been no mind-bending experiments – apart from the CIA's notorious Project MK-Ultra (see below) – Hunter was simply weaving a web of lies. In truth, the Chinese did run re-educational programmes for any prisoner who wanted to attend. Referred to as *szu-hsiang-kai-tsao*, or thought reform after mind cleansing, these programmes were an attempt to rid the PoWs' minds of their pre-conceived notions of Communist China, and replace those 'Western lies' with 'the truth'. There were no drugs, no hypnotists, no beatings; just boring lessons. That said, over 2,000 PoWs refused repatriation to America after the war, and America therefore needed to believe they had been induced to do so by sinister means rather than by choice.

A SINISTER TURN

Perhaps unsurprisingly, the CIA too was more than a little interested in Vicary's experiments. In their post-war, clandestine Operation Paperclip, the CIA had spirited back to America numerous Nazi scientists and doctors, some of whom had used death-camp inmates in various psychotropic and sensory-privation experiments in their own attempts to achieve total mind control. More than thirty war criminals were given new identities and employed in the set-up of the decidedly sinister MK-Ultra programme, which used unwitting members of the public in a series of dangerous and, in some cases, lethal psychotropic experiments.

The programme began to flag by 1957, but it seems that Vicary unwittingly breathed new life into it. A recently released CIA report dated 17 January 1958 conjectured:

> It might be that in order to lessen the resistance of an individual to the hypnotic induction process, the use of subliminal projection may be considered. This technique has achieved success in commercial advertising, as 'eat popcorn' or 'drink Coke' projected in certain movie theaters for 1/3000-of-a-second intervals. It may be that subliminal projection can also be utilized in such a way as to feature a visual suggestion such as 'Obey [deleted]', or 'Obey [deleted]' – with similar success.

THE LID IS LIFTED

The 1952 production of an animated version of George Orwell's *Animal Farm* had been a CIA-funded initiative and flash-messaged re-screenings were organized, but the experiment failed to yield any results. And while the CIA remained in a state of confused frustration in their attempts to manipulate whole cinema audiences into a certain frame of mind, or to undertake certain actions immediately on leaving the auditoria, American and Canadian broadcasting networks were equally unsuccessful in their own attempts to replicate Vicary's results.

The most publicized of these independent experiments was carried out by the Canadian Broadcasting Corporation during its much-viewed Sunday night show *Close Up*. During the programme, invitations to the viewers to phone in to the station were flashed on the screen nearly 400 times, but the station did not receive a single call.

The next nail in the subliminal-message coffin was hammered home by Dr Henry C. Link, Director of the Psychological Corporation. He invited Vicary to demonstrate the supposedly incredible power of his technique under controlled conditions, with Vicary failing on all fronts. Then, in 1958, Stuart Rogers, a student of psychology at New York's Hofstra University, visited Fort Lee, the home of Vicary's original experiment, to ask some very direct questions that had not yet been asked. When Rogers spotted the small local cinema where Vicary had carried out his experiment he was immediately struck by its size – it was much too small to have accommodated the numbers claimed by Vicary over the indicated time span. After an interrogation from Rogers, the manager of the cinema admitted that no such experiments had taken place.

CARRYING ON REGARDLESS

Finally, in 1962 Vicary himself admitted that it had been a scam: there was no such thing as subliminal messaging; he had invented it all in an attempt to save his failing consultancy. Despite these revelations, the bandwagon Vicary had so deftly put in motion refused to listen.

No matter how loud or how often Vicary protested that subliminal messaging was nothing more than a scam, the world failed to listen. Studies conducted in 2006 show that in America today, over 80 per cent of people – including those who work in advertising or teach psychology – still believe in the sinister power of subliminal messaging. Current opinion was certainly bolstered by the American author Wilson Bryan Key (1925–2008), who during the 1970s and the 1980s became the self-styled guardian against such evils.

Despite impressive academic credentials, including a PhD in communications and a membership to American Mensa, Key was a firm believer in the power of subliminal messaging.

THE ANGELS OF MONS

As the case of James Vicary suggests, once a nation has taken an idea to heart, it becomes very difficult to persuade its members otherwise. Much the same happened to the British writer Arthur Machen, whose creations spawned the legend of the Angels of Mons. The entirely fictitious story described how a host of angels protected British soldiers on the WW1 battlefields, holding back the gibbering Hun with flaming swords and arrows. Although it was only a story, hordes of troops, officers and chaplains, both Allied and German, claimed to have witnessed the event for themselves. Machen's reward for trying to remind the nation that it was but a whim of his own invention was a horse-whipping from an outraged bishop who collared him in Oxford Street.

A cover design for solo piano music inspired by the myth of Mons

He claimed that three of the ice cubes in a Gilbey's Gin advert contained the letters S, E and X discernible within their structure; he even saw the same word if the perforations along the edge of Ritz Crackers were joined together. Key stood watch for the Christian Right lobby, forcing the multinational product company Proctor and Gamble to abandon its logo of a grey-bearded man posing as a moon because he claimed the figure '666', the Number of the Beast (see box below), was discernible in his facial hair.

666?

Did you know the Number of the Beast was and still remains 616, not 666? Modern translators of the Bible, makers of popular Satanic films and thousands of Goth rockers have been misled.

ROCK 'N' ROLL SUICIDE

But still the corrupt legacy of James Vicary rolled on. Buoyed by the success of their farcical crusade against Proctor and Gamble, the subliminal watchdogs turned their attention to Gothic-themed heavy rock groups. As far as Wilson Key was concerned, there had to be subliminal hooks somewhere in the equation to help explain the attraction of their blasphemous decadence. He reasoned that there must be a call to join the ranks of Satan in the words when played backwards (backmasking). Naturally, everyone started playing their record collections backwards, and the Christian Right believed this to have been the cause behind many teenage suicides, after the victims received 'subliminal' orders to take their own lives.

So, according to Key and his like, not only was the brain able to see images that normally escape the human eye, it could also listen to a record played normally, log the entire performance and then play it in reverse in order to pick up any hidden subliminal message that may have been lurking in the lyrics. Quite a mental feat. Basically, no one wanted to accept the obvious: teenagers who filled themselves up with drugs and were given to the darkness of goth-rock culture might simply be predisposed to self-harm. But matters came to a head with the suicide of American teenager Raymond Belknap and the attempted suicide of his friend James Vance on 23 December 1985.

Both young men had a history of drug abuse and depression and had allegedly spent the day smoking marijuana and listening to hours of music by the British heavy metal band Judas Priest. They then wandered off to a graveyard where Belknap blew his head off with a shotgun leaving Vance to try the same; he survived but with horrific injuries. Wilson Key was wheeled out as a professional witness on the power of subliminal messaging for a trial in which the rock group was charged with the notion that their track 'Better By You, Better Than Me' carried embedded subliminal messages which had urged the boys to take their tragic actions.

Fortunately, the judge was not that gullible and, with Key's pronouncements discounted, the case was dismissed. As Judas Priest's lead singer, Rob Halford, would wryly observe after the case, 'If I'd thought that sort of crap worked, we would have embedded messages telling everyone to go out and buy more of our records.'

SPURIOUS SPIN-OFFS

The gullibility of the public at large is indeed an awesome market force. One might be forgiven for assuming that if premise B is built on premise A, which is later proved to be a complete lie, then premise B would automatically go into the bin of history, along with its progenitor. Not so. Despite Vicary's very public admission that subliminal messaging had been a lie, spurious spin-offs had already been set well in motion.

Although the idea of sleep-learning had been quashed in 1956 with the electroencephalography studies conducted by Charles Simon and William H. Emmons of the Rand Corporation, Vicary's lie breathed new life into the notion. Commercial ventures sprang up across America and Europe, trumpeting the false promise of allowing us to tap into the fearsome powers hidden in the subconscious – if subliminal influence could exert such power when the subject was awake, its power would be even greater if the subject was asleep. It is just a pity there is no truth in the idea.

A multi-billion-pound industry has been built on the idea of sleep-learning, which – despite having been roundly debunked by prominent sleep experts and psychologists – still refuses to go away. In studies conducted in 1991 participants were told that personal enhancement messages would be played to them while they slept to help make them more assertive. On waking, many participants did indeed proclaim themselves to feel more in control and behaved accordingly – despite having been subjected to messages telling them to be more humble and self-deprecating. In all such experiments, conducted under clinical conditions and with the subjects wired up to an ECG to ensure they really were asleep, the knowledge impartation has been nil.

THE MOZART EFFECT

From subliminal messaging and sleep-learning stemmed the 'Mozart Effect', a term coined in 1991 by the French ENT specialist Dr Alfred A. Tomatis (1920–2001). Tomatis proclaimed that listening to certain types of music was beneficial to certain conditions, and conducive to the achievement of certain objectives. Among the many types of music Tomatis recommended, Mozart was thought to aid depression and concentrate the mind of those with learning difficulties to the task at hand.

Such was Tomatis' international standing, two University of California heavyweights, Dr Frances Rauscher and Dr Gordon Shaw, looked into the alleged phenomenon and published their results in the scientific journal *Nature.* For reasons only they could conjecture – perhaps concentrating on the music helped the mind to warm up – subjects who listened to Mozart before certain tests did indeed do better than those who had been left in the dark in silence.

The good doctors said they had detected a slight and very transient improvement in participant's spatial-temporal reasoning powers and that 'there appear to be pre-existing sites in the brain that respond to specific frequencies'. They made no mention of individual participants' IQs, nor that other experiments had reported similar findings using the music of Meatloaf and Iron Maiden. Predictably the press reported that listening to Mozart makes your children smarter – the so-called Mozart effect was ripe for exploitation.

Despite Shaw and Rauscher protesting that their work had been completely misrepresented – that there was no

WELL I NEVER! POPULAR SCIENTIFIC IDEAS DEBUNKED

- We use all of our brains (well, at least some of us do!) and not just the 10 per cent of myth.
- There is no harm in waking up a sleepwalker; in fact, it is best to do so.
- Leprosy cannot be transmitted by casual contact.
- 20/20 vision does not denote perfect eyesight, only that both eyes function normally at a 20ft range.

correlation between the playing of Mozart and the listener's intelligence – the idea had already been set in motion. In 1998 the Governors of both Georgia and Tennessee announced budgets to provide every new-born baby with music CDs, and new research was conducted into the awareness of the unborn child.

Millions of mothers-to-be snapped up devices that comprised a CD player and an implement that resembled a reverse-stethoscope to bombard their unborn with Mozart, along with recordings of their own voices making positive affirmations of life. However, Georgia and Tennessee failed to become the cradles of *wunderkind* and we have yet to read of the first child to launch into a rendition of *Don Giovanni* as the midwife slaps it into life.

Victoria's Secret

Cocaine and heroin can cure a range of man's ills

LONG BEFORE THEY were recognized as Class-A drugs, capable of destroying the lives of those who grow, import and consume them, cocaine and opium were hailed as panaceas by the nineteenth-century medical profession. Eulogized as cures for a range of ills, the drugs were made readily available in numerous over-the-counter products, many of which were sponsored by leading doctors and prominent figures.

THE TREND CATCHES ON

Pope Leo XIII (1810–1903) never ventured far without a hip flask of the cocaine-laced French wine Vin Mariani; he even bestowed the Vatican Gold Medal on the manufacturers and allowed his image to be used in the promotion of the product, which was marketed at ailing children, pregnant women or those feeling just a little under the weather. At that time, everyone thought opium and cocaine to be of great benefit to mankind at large.

Vin Mariani: the Pope's favourite

Mothers were also encouraged to use either drug to treat teething babies. Since the sixteenth century, misguided physicians had believed in the efficacy of slitting the gums of infants to help ease the emergence of the first teeth. The practice had gone into decline by the end of the eighteenth century, mainly for the distress it caused, but a century on it enjoyed a comeback as the pain of the incision could be followed with a direct application of one of these new 'miracle anaesthetics', after which no doubt the child just sat there chortling.

With the medical profession trumpeting their merits, not to mention profiting from their mass production and sale, cocaine- and opium-based products were readily available; even greengrocers and haberdasheries kept stocks for anyone caught short. And the hallmark of cynical advertising was all-pervasive. All such drugs were recommended by an avuncular-looking Harley Street consultant, sold under innocuous-sounding names such as Mrs Winslow's Soothing Syrup or Professor MacGuire's Infant's Panacea. The entire medical profession was united in the opinion that opiates and cocaine were valuable cures that contained no addictive qualities. But because the medical professionals paid to investigate the nature of the two drugs were in fact voracious consumers themselves, it was little wonder that their findings completely exonerated the drugs of any charges regarding their capacity to addict.

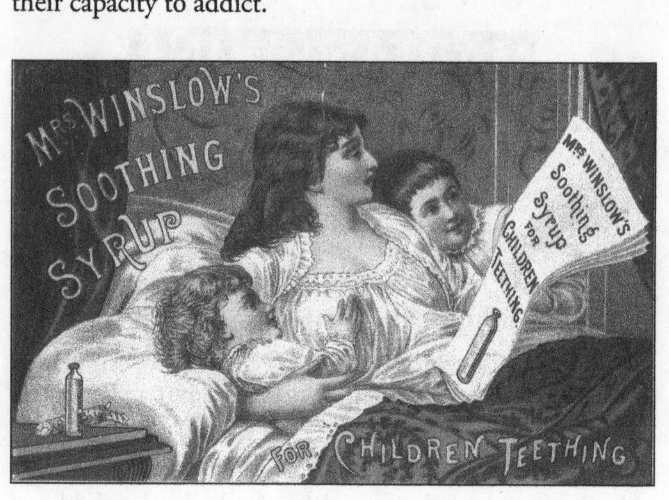

Soothing syrup for children: Mrs Winslow gives her seal of approval

IF FREUD SAID IT'S OK ...

The Austrian neurologist Sigmund Freud (1856–1939) championed cocaine not only as the ideal cure for opium and alcohol addiction, but also as an appetite stimulant for those with anorexia, and as a first-line treatment for asthma sufferers.

In *Über Coca* (1884) Freud applauded the drug's efficacy in stimulating 'exhilaration and lasting euphoria, which in no way differs from the normal euphoria of the healthy person'. Warming to his theme, Freud continued:

> You perceive an increase of self-control and possess more vitality and capacity for work. In other words, you are simply normal, and it is soon hard to believe you are under the influence of any drug.
>
> Long intensive physical work is performed without any fatigue. This result is enjoyed without any of the unpleasant after-effects that follow exhilaration brought about by alcohol. Absolutely no craving for the further use of cocaine appears after the first, or even after repeated taking of the drug.

BEST OF BRITISH

Queen Victoria, on her physicians' advice, guzzled laudanum, a whole-opium tincture; she took cannabis for her menstrual pains; and, like her Prime Minister, William Gladstone, she also liked the occasional snort of cocaine. It is probably no exaggeration to say that, throughout the greater part of the nineteenth century, the British ruling class and half of the population lived in a drug fug. And why not? The government in Victoria's day was the most aggressive drug cartel the world has yet seen; they even made the Columbian warlords look like amateurs. The UK's domestic imports of raw opium in 1830 had been alarming enough – just short of 100,000 1lb blocks; by 1860 this had rocketed to just under 300,000lb.

It was the British who so dramatically expanded the poppy fields of Northern India and Afghanistan, the deadly results of which still reverberate today. Desperate to find new markets, the British Empire targeted China, despite the country's already serious opiate problem, with a total of approximately 2 million addicts. Regardless, taking advantage of a great opportunity, the British began shipping great quantities of the drug to expand that user base. When the Chinese objected and demanded the trade cease, the good Victorians attacked Canton, the means by which China controlled trade with the West, in the First Opium War (1839–42).

MEDICINE MEN: WHEN DOCTORS GOT IT WRONG

The following list of dubious medical procedures suggests the men of medicine haven't always got it right.

- Trepanation (you'd need it like a hole in the head!) to cure all manner of ills.
- Lobotomies to 'calm' the mentally ill.
- Electroconvulsive therapy to treat depression.
- An insulin-induced coma to relieve the symptoms of schizophrenia.
- Hysterectomies to cure paranoia in women.

The British emerged victorious and demanded from China a host of concessions in the Treaty of Nanking, which included ceding Hong Kong to Queen Victoria. The Chinese were reluctant to adhere to all of the treaty's terms, leading the British to initiate another attack in the Second Opium War (1856–60). The British again emerged victorious. They added Kowloon to their previous acquisition, forced the Chinese to legalize an opium trade that was destroying their people, and began the wholesale transportation of 'indentured Chinese labourers' (slaves, in other words) to the Americas to toil on the railways. China had no choice and within a matter of years the British government had acquired over 100 million opium customers in China alone.

HOUSTON, WE HAVE A PROBLEM

The situation in America was no better, where a staggering 10 million opium pills and over 3 million ounces of other opiate preparations were dished out to the Union Army alone during the Civil War (1861–65). While the drugs may have helped the troops to get through the horrors of war, when the conflict was over and nearly 500,000 opium-habituated troops returned to civilian life, the American medical profession was forced to acknowledge that the unrestricted use of such drugs might be something of a problem.

Those who could afford to fund their habit could buy the drugs locally over the counter, or through mail-order from medical suppliers who still upheld the common belief that these were harmless substances, even if put to recreational use. It was the antics of those who could *not* afford their newly acquired habit that first gave cause for concern.

Backed by the American Medical Association, the Society for the Suppression of the Opium Trade renewed its attack on doctors who considered opiates and cocaine to be not only harmless, but also broadly beneficial. In Britain in 1893 a Royal Commission on Opium was reluctantly convened to investigate. But this was a venture doomed from the start. Although the commission went through the correct channels, listened attentively to countless witnesses, and produced a suitably weighty report, it concluded that the use of opium was not responsible for a decline in moral standards or physical damage to the user. In fact, it decreed that opium in recreational use was no worse than alcohol, and that it had positive medical benefits.

The pre-eminent British medical journal *The Lancet* welcomed the findings, commenting that they 'dealt a crushing blow to the anti-opium faddists whose claims were either ridiculously exaggerated or wholly unfounded'. The problem with the commission was that it was a *Royal* Commission and, as such, directly answerable to the junkie-in-chief herself – and no one involved relished the prospect of confronting Queen Victoria with the realities of her own laudanum and cocaine usage.

DENS OF INIQUITY

Novels promoted the myth of opium dens run by Chinese immigrants, who supposedly teemed in London's Limehouse district. While the Chinese certainly ran opium parlours in American cities, to where they had been transported following the Second Opium War, they never numbered more than a few hundred in nineteenth-century London. There is in fact no evidence of any such dens. Why go to a seedy den in a dangerous part of London when you could buy your stash in Mayfair?

STRAIGHT TO USER

Literature began to explore the seedy and debilitating side of opium, with Conan Doyle's Dr Watson expressing

An artistic representation of one of London's fabled opium dens

serious disapproval of Sherlock Holmes' use of a substance deleterious to his superior intellectual powers.

Few wanted to listen to the dissenting voices on the ill effects of opiates and cocaine, and the medical profession vehemently defended its stance. In 1885 the American pharmaceutical company Parke-Davis augmented its cocaine product range with a solution that could be injected directly into the user's veins. It was sold in a variety of high street shops, complete with syringe, and backed by an advertising campaign that claimed it would 'supplant the need for food, make cowards brave, make the timid loquacious and render the user insensitive to pain.'

SERVING THE MASS MARKET

In 1898 the German pharmaceutical house Bayer synthesized a new and highly addictive opoid, which they marketed as an over-the-counter product under the brand name of 'Heroin', for the supposed heroic feelings it engendered in the user. Again, the medical profession welcomed this advance without question, and advocated the use of heroin for the

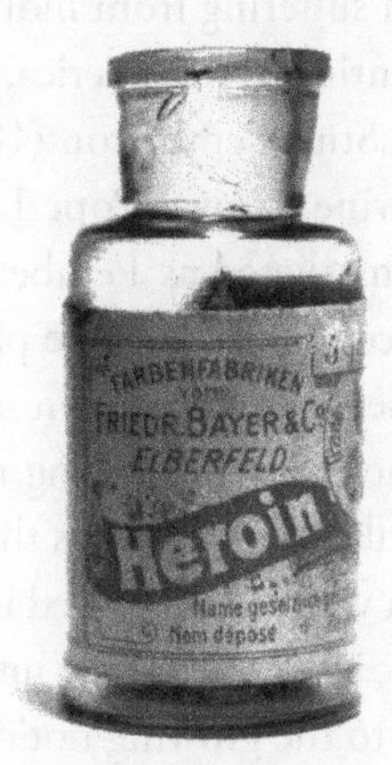

Over-the-counter: one of German pharmaceutical house Bayer's heroin products

More than just a thirst quencher: Coca Cola was advertised as a stimulant to help relieve exhaustion

treatment of colds and flu, bronchitis, whooping cough, and for pregnant women suffering from morning sickness.

Developments continued in America, where Confederate veteran doctor John Stith Pemberton (1831–88) invented a new cocaine-laced wine to rival Pope Leo XIII's favourite, Vin Mariani. First marketed as Pemberton's Brain Tonic, this lethal mix of alcohol and cocaine proved very popular, until in 1886 Pemberton's hometown of Atlanta, Georgia, objected to its alcohol content, forcing the doctor to create a Temperance-friendly version. It was this blend of cocaine-laced syrup and soda water that evolved into Coca-Cola. But the manufacturer became increasingly uneasy about the drug element in response to the growing notion in the South that cocaine usage was driving black men to rape white women. In 1903 all traces of cocaine had been stripped from the drink.

NEW BOY IN TOWN

The writing was by then on the wall for the continued free availability of cocaine and heroin. It had dawned on the majority of the medical profession that the unchecked use of these drugs was not the answer to the people's minor ailments. That said, it was not until 1920 that possession of opiates became illegal in the UK, and the possession and use of cocaine remained unlegislated in America until 1970.

Meanwhile, by the 1950s the Americans had taken to their hearts another 'harmless' drug – amphetamine. Doctors on both sides of the Atlantic scattered these pills like confetti. Apparently they found them very useful to perk up patients in hospital, and everyone was buying Benzedrine inhalers over the counter. Throughout the 1950s, Pan-Am's in-flight courtesy packs for passengers included a Benzedrine inhaler, which the accompanying pamphlet advised 'would make the flight more pleasant and seem to pass more quickly'. No kidding!

Heaven Scent

Disease is caused by foul smells and lack of personal hygiene

Known by many as the father of germ theory, the French chemist and microbiologist Louis Pasteur (1822–95) conducted a series of experiments in the second half of the nineteenth century that conclusively proved germs cause disease. Before Pasteur's discoveries, the overwhelming majority of the medical and scientific communities, including pioneering nurse Florence Nightingale (1820–1910), had publicly ridiculed the idea that organisms undetectable to the human eye could invade the body, breed in sufficient numbers and bring about the death of the host.

KICKING UP A STINK

Until Pasteur's irrefutable evidence was published, the 'miasma theory' held sway. This concept maintained that all disease was caused by foul smells, lack of personal hygiene and, to a lesser extent, impurity of the mind and soul. Florence Nightingale certainly adhered to this notion.

Her complicity may have caused thousands of deaths in the Crimea, where she thought that clean surroundings and regular Bible readings would secure a patient's road to recovery.

BY THE POWER OF GOD ...

Before the miasma theory gained ground, the majority of the Ancient civilizations believed disease to be a punishment inflicted by the gods. Although detractors did exist: *Atharvaveda*, the sacred Hindu text written towards the end of the second millennium BC, suggested living causative agents were responsible for disease; the text did not elaborate on the size and nature of the agents, so it may have been nothing more than guesswork.

Marcus Terentius Varro (116–27 BC), a Roman physician and polymath, was the next to suggest something more than bad air and foul smells might be responsible for disease. Not only did he caution people against building homes near swamps, he even suggested they refrain from frequenting them, because proliferating therein were 'certain minute creatures that cannot be seen by the eyes, which float in the air and enter the body through the mouth and nose and there cause serious diseases.' But few paid attention to Marcus Terentius; majority opinion held the gods to be responsible.

It was not until the sixteenth century that the infection theory was further advanced. The Italian physician Girolamo Fracastoro (1478–1553) believed that infective agents called spores were responsible for the spread of germs and viruses; he warned that these spores could transfer infection by direct or indirect means. He also wrote that 'such things as clothes, linen, etc., which although not themselves corrupt, can nevertheless foster the essential seeds of the contagion and thus cause infection.' Although Fracastoro was on the right track, it was to be approximately another 300 years before the Italian physician Agostino Bassi (1773–1856) made the first identification of living organisms as the cause of disease.

GERMINATION

In 1835 Italian silk production was in danger of collapse because the silkworms had fallen prey to an invasion of parasitic mites. Bassi made note of the white, powdery coating of spores found on the dead and dying silkworms, and he became the first to make the correct connection between infective invasion and disease. Thirty years later, when the French silk industry fell prey to the same parasite, Louis Pasteur, inspired by the work of Bassi, came to the same conclusion although, in Pasteur's defence, he had by that point already made major inroads into the discovery of germ theory.

As Bassi had before him, Pasteur recommended the separation of the colonies of silkworms into previously disinfected farms and the immediate destruction of any silkworm that showed signs of infection. Despite these quarantines and precautions having twice eradicated the epidemic, the medical and scientific communities were not prepared to listen. The majority of opinion still held that all disease – cholera, typhoid, malaria, to name but three – was the inevitable progeny of foul smell. In Austria, Pasteur's contemporary, Ignaz Semmelweis (1818–65), was also leading a case against miasma theory, but so vehement were his attacks that it would eventually lead to his murder by a cabal of some of the most prominent medical figures in Austrian medicine.

Le Petit Journal

LE CHOLÉRA

This front page of a French newspaper from 1912 depicts Death carrying cholera

PLEASE WASH YOUR HANDS

Death rates of approximately 20 per cent for mothers and offspring during childbirth were considered quite normal in some clinics in the mid-nineteenth century. This included Vienna hospital's First Obstetrical Clinic, which offered benefits to expectant mothers if they volunteered themselves as guinea pigs for new doctors.

There were two maternity clinics at the hospital. Semmelweis took up his position as Senior Resident at the First Obstetrical Clinic in1846, but was puzzled to find that the childbirth death rate at the Second Clinic was a mere 2 per cent – 18 per cent lower than the First Clinic. The only difference between the two institutions was that no autopsies were conducted at the Second Clinic. His second clue came from the patients themselves. Having noticed that something was seriously awry at the First Clinic, a goodly number of Vienna's pregnant women contrived to give birth en route to the First Clinic in order that they remain entitled to the benefits promised by the programme without having to run the gauntlet of its professional care.

Ignaz Semmelweis with mother and child

Semmelweis was staggered to find that the death rate in this forward-thinking group was almost non-existent. Semmelweis' predictions were confirmed with the sudden death of his good friend Professor of Forensics Jakob Kolletschka in 1847. Kolletschka had been supervising a student conducting an autopsy on a patient who had died of puerperal (childbirth) fever in the First Clinic and, within three days of being nicked by a carelessly brandished scalpel, he was himself dead of the same condition.

Focusing his attentions on the autopsy rooms, the first thing Semmelweis noted was that students and professors routinely left them without washing, and headed straight to the treatment and examination rooms. He immediately instituted a regime of hand-washing in a solution of

WELL I NEVER! POPULAR SCIENTIFIC IDEAS DEBUNKED

- Edison did not invent the light bulb.
- Benjamin Franklin did not invent the so-called Franklin stove.
- Pythagoras did not come up with 'his' eponymous theorem.
- The Wright brothers were not the first to achieve powered flight.
- Alexander Bell was not responsible for inventing the telephone.

chlorinated lime, and the death rate fell by 90 per cent overnight. Within two months it was nil. However, instead of being hailed a hero, Semmelweis was hounded out of his post by colleagues offended at being ordered to wash their hands like naughty schoolboys. What possible correlation could there have been between the hand-washing fiasco and the fall in the death rate?

THE COLD SHOULDER

When Semmelweis left the clinic, the death rate returned to its original position. In every subsequent post he took up, Semmelweis instituted the same hand-washing regime and the death rate plummeted, but his old enemies had him manoeuvred out of every position he assumed – and the death rate rose in his wake. And still nobody listened. Unbelievably, the Viennese Medical Council saw Semmelweis as a troublemaker, intent on blaming patient and infant death on some of its more influential members, who felt it their right to poke around inside patients with hands still dripping with germs.

Roundly ridiculed as a fractious crank, Semmelweis was cut out of the medical fraternity and marginalized. But he remained unrepentant. Continuing his campaign in exile, Semmelweis published an open letter to the obstetrics fraternity at large, which was, unfortunately for its sender, a little on the antagonistic side. Feeling that decisive action was needed, Semmelweis' enemies decided to have him declared insane and locked away where he could create no more waves.

In 1865 a small group of physicians, led by Vienna's leading dermatologist Ferdinand Ritter von Hebra (1816–80), lured Semmelweis to an asylum under the pretence that his opinion was sought on certain matters. No sooner was Semmelweis through the door, he became suspicious and tried to leave, but von Hebra's henchmen were prepared and they beat Semmelweis so badly that he died shortly after in one of the asylum's dungeons.

From an early twentieth century book on the new rules of hygiene. The bed on the right was thought to trap germs

It is quite understandable that such entrenched and established opinion would be a large ship to turn; but when the undisputable proof was clear for all to observe, it seems bizarre in the extreme that the only action the captains of that ship could think to take was to murder the messenger. But Semmelweis is not forgotten. In scientific circles, the 'Semmelweis reflex' is a common expression used to describe the knee-jerk reaction of entrenched thinking to any suggestion that it might be wrong.

The Origin of the Specious

There is a missing link in the evolutionary chain

THE NOTION THAT there was a missing link in the evolutionary chain, an early human ancestor that would help explain the full transition from ape to man, was an accepted idea of both palaeontology and anthropology until the early part of the twentieth century.

The descent of man: a smooth transition that bore little relation to reality

THE SEARCH BEGINS

Although the concept has no basis in scientific fact, the search for the missing link among the archaeological community has persisted since the late nineteenth century. While many have sought proof of such a link, others believed they had found it. A contingent of nineteenth-century anthropologists believed it still stalked the more remote regions of the planet; one demented soul even tried to breed it, with consequences that still ravage the Third World to this day.

The Dutch palaeontologist Eugène Dubois (1858–1940) was the first to embark on a planned search for evidence of a missing link. In 1890 he was to be found digging up half of Java, Indonesia in his quest. In the latter part of that year he dug a pit near the Solo River in Eastern Java and unearthed the remains of what would become known as 'Java Man', an admittedly simian-like creature that Dubois immediately

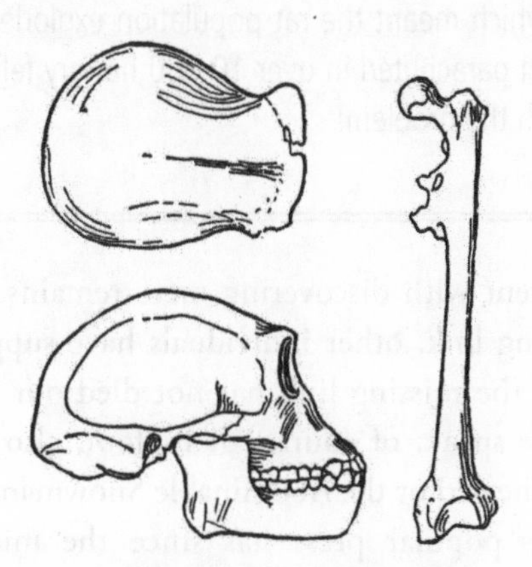

The remains of 'Java Man'

pronounced to *be* the missing link. Few believed him and he drew so much ire from academics that he gathered up the bones and became a virtual recluse. Dubois finally allowed full examination of Java Man in 1923, when it was quickly established that his find was nothing more than an example of *Homo erectus* (upright man). Dubois retreated into obscurity, never again to communicate with academe.

IT FELT LIKE A GOOD IDEA AT THE TIME

In the 1950s the WHO decided to combat Borneo's malaria problem by killing mosquitoes with DDT spray. Although a success, the spraying also killed off the wasps that fed on the roof-munching thatch-caterpillar, which promptly de-roofed the homes in the local area. Lots of cats also died, after trying to lick themselves clean, which meant the rat population exploded. The RAF then parachuted in over 10,000 hungry felines to deal with the problem!

Not content with discovering mere remains of the so-called missing link, other individuals have supported the notion that the missing link has not died out but is still with us. We speak, of course, of Bigfoot, also known as Sasquatch, the yeti or the Abominable Snowman.

America's popular press has since the mid- to late nineteenth century peddled the notion of Bigfoot as a living missing link. A rather surprising coterie of academic heavyweights have stood firmly in this corner, including Grover Krantz (1931–2002), Professor of Anthropology at Washington State University, and his colleague Geoffrey Bourne (1909–88), Director of the internationally respected Yerkes National Primate Research Center of Atlanta, Georgia.

Such a notion was supported by the UK's John Napier (1917–87), an anthropologist, primatologist and leading light at America's Smithsonian Institute. The celebrity anthropologist Margaret Mead (1901–78) also believed that hairy missing links still stalked the snowlines – however, she also believed in UFOs and guardian aliens who were sent here to observe us and ensure civilization was kept on the straight and narrow.

In 1953 the explorer Sir Edmund Hillary reported seeing yeti tracks and mounted an expedition in 1960 to gather evidence of the beast's existence. In 1959 actor James Stewart took possession of a supposed skeletonized yeti hand that had been stolen from Tibetan monks; Stewart smuggled it out of India to London for examination, where primatologist William Charles Osmond established that the hand was in fact Neanderthal. The American anthropologist Jeffrey Meldrum (1958–), Professor at Idaho State University, actually went in search of Bigfoot in Siberia in 2011, although his expedition failed to yield any results.

JUST ONE PROBLEM

The problem is these investigations had no basis in scientific fact: no one, Darwin included, has ever suggested that man is descended from apes. Had it been the case then there would be no apes left – they too would have evolved and would doubtless now be running banks, or taking up their seats in the House of Commons. And despite nineteenth-century academe presenting the concept of man's descent in linear form, with clearly defined stages of development and advancement, this is now known to have been very wide of the mark.

THE DESCENT OF DOG

Many species, including cats and dogs, were once the same species with common roots in the prehistoric miacis, an arboreal dog-like creature with retractable claws used for climbing. This ancient progenitor subdivided by dictate of its diverse environments to produce cats, dogs, weasels, bears and hyenas. Yet no sensible person would propose that dogs were descended from cats. It is also worth mentioning here that the hyena is the very confused marker of the cat– dog divide because, for all its appearance and countless references to it as a pack-hunting dog, the hyena is in fact a cat.

The different skeletons in profile. (From left to right) gibbon, orangutan, chimpanzee, gorilla and man

Desmond Morris' book *The Naked Ape* (1967), which sought to compare humans to other animals, is partially responsible for suggesting man descends from apes, as are the ever-popular posters that depict the evolution of man as a series of figures that become increasingly upright and devoid of body hair. What Darwin proposed – as others had before him – was that man and primate shared a common progenitor before dividing to descend in parallel, rather than in sequence.

The traditional perception of the descent of man as a sequential series of neat changes is completely inaccurate. The slow and painful development of arguably the planet's most dangerous animal followed a multi-branched and rather untidy path, with some branches dying out and others co-existing and even interbreeding across millennia. There were no cleanly divided stages.

LENGTHY DIVERGENCE

In 2006 the exploration of the complex history of the human genome conducted by David Reich, a population geneticist at Harvard, revealed that the divergence of proto-humans and proto-chimpanzees from their common progenitor was a more lengthy and complicated process than previously thought. Just as late Neanderthals and early modern man co-habited and interbred (see box opposite), so too did proto-humans and proto-chimpanzees over several million years. The recent tracking of the X chromosome in both modern man and chimpanzees reveals that the final divergence of the two species did not actually occur until between 5 and 6 million years ago, approximately 1 million years later than previously thought. But this still does not give support to the spurious theory of a missing link; it just means that early humans and chimpanzees remained similar enough in appearance for longer than was previously assumed, after dividing from the common progenitor, to find each other acceptable breeding partners.

THE IDEA WILL NOT DIE

Despite the fact there has never existed a tidy chain in our evolution from which an irksome missing link could have absconded, the idea would not die. Creationists, driven to a frenzy of denial by the publication of Darwin's *On the Origin of the Species* (1859), pilloried him for allegedly suggesting that man was evolved from apes rather than created by God.

NOT SO NEANDERTHAL

After the remains of a Neanderthal man were found in 1856 in the Neander Valley, just to the east of Düsseldorf, academics rushed to brand him a brutal, hunched and hairy club-wielding humanoid. In reality, the Neanderthal was a far more civilized being.

Although slightly shorter and stockier than his modern counterpart, Neanderthal man was no troglodytic dolt. Possessing a brain approximately 100ml *larger* than his modern counterpart, Neanderthal man built individual and group dwellings, and kept himself warm with fires. He cooked meat and vegetables, had his own language, and produced tools as sophisticated as those of co-evolving modern man, with whom Neanderthal man socially interacted and crossbred, as the present gene pool proves. (Approximately 4 per cent of the DNA in Europeans and Asians is Neanderthal.)

As for the once-held belief that 'primitive' Neanderthals were driven to their rightful extinction by the rise of so-called modern man, this too has been quashed. Advances in DNA indicate that Neanderthal man was simply absorbed into early Cro-Magnon stock and modified through interbreeding. Neanderthal man, therefore, did not die out – he is still with us, a fact that will come as no surprise to female readers.

As previously stated (see From Mendel to Mengele on page 57), Darwin never suggested anything of the kind, but the press, among others, decided this was the thrust of his work; and those who had never read Darwin's seminal book were so keen to attack they leapt on this ready-to-ride bandwagon.

By the advent of the 1900s, a British amateur archaeologist called Charles Dawson (1864–1916) was making a name for himself after he dug up half of Sussex to unearth some pretty interesting finds. In 1912 Dawson dramatically announced a staggering discovery: he had found the missing link – the skeletal remains of the long-sought ape-man, near the village of Piltdown, East Sussex.

It would be over forty years before the 'discovery' would be denounced as a fraud and the bones revealed to be the composite of a medieval human skull, the lower jaw of an orang-utan, and chimpanzee teeth, all filed down to resemble a more human profile, and appropriately aged in chemical solutions. But whether Dawson himself was the fraudster, or had been unwittingly duped by another who had placed the find in his path, has never been satisfactorily answered. It is, however, interesting to note that a near neighbour was Sir Arthur Conan Doyle, a man who was himself under heavy fire from the Christian Fundamentalists for his championing of the spiritualist movement. Not only was Conan Doyle a resident of Piltdown, he was also, like Dawson, a member of the Sussex Archaeological Society; the two would often be seen huddled together deep in conversation in the run-up to the 'discovery'.

In 1953 *Time* magazine published an article that completely debunked the so-called Piltdown Man, but by

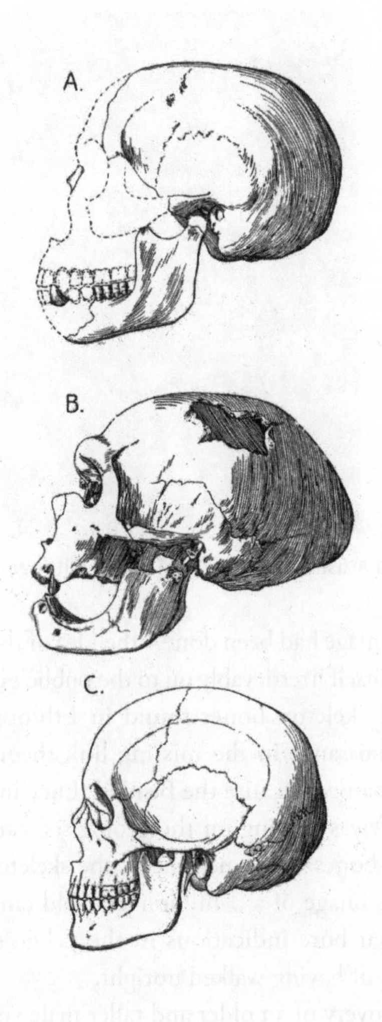

Compare and contrast: the Piltdown skull (A),
a Neanderthal skull (B) and a modern human skull (C)

An artistic reconstruction of the Piltdown man

then the damage had been done – the idea of the missing link had etched itself irretrievably on to the public's consciousness.

In 1974 skeleton bones found in Ethiopia heralded a further renaissance in the missing link theory. Known as 'Lucy' (so named because the Beatles' 'Lucy in the Sky with Diamonds' was playing on the geologists' camp tape deck as the first bones were unearthed), the skeleton did indeed present the image of a 3-million-year-old chimpanzee-like creature that bore indications in the pelvic structure and knee-joints of having walked upright.

The discovery of an older and taller male version of Lucy in Ethiopia in 2010 lead to further excitement within some circles about the prospect of a missing link. But no one involved in either dig had at any point suggested the finds were anything of the sort. Imagine for a moment that most traces of present humanity were removed from this planet, and archaeologists from a future alien planet unearthed the remains of John 'Elephant Man' Merrick and the diminutive Major Tom Thumb from Barnham's Circus. If they tried to build their perception of our civilization on those two finds, how wide of the mark would they be? Perhaps there did exist isolated groups of long-since-extinct apes who chose to walk upright but were wiped out for their audacity – it still does not place them in the line of man's descent.

STAGED INTERVENTION

And so to the man who decided to produce his own missing link – Stalin's own Dr Frankenstein, Soviet biologist Ilya Ivanovich Ivanov (1870–1932) and his Humanzee Project of the late 1920s. As early as 1910, Ivanov was openly musing the feasibility of creating a living, breathing missing link by cross-breeding humans and chimpanzees, orang-utans and even gorillas. Few doubted him; this, after all, was the man who had already produced a zeedonk (a zebra crossed with a donkey), and cross-bred or cross-inseminated a range of animals from cows with antelopes to rabbits with rats.

By 1925 Ivanov had secured Humanzee funding from the Kremlin and, rightly thinking such experiments to be better conducted as far away from Moscow as possible, he and his son, also called Ilya, headed for the less stringently controlled environs of Africa. On a more sinister note, Ivanov considered Africa best suited his programme because, as 'everybody knew', Africans are closer to our simian forebears than are white Europeans, and he was going to need human participants in his grotesque fiasco.

The summer of 1926 saw the Ivanovs ensconced in Conakry, in the former French Guinea. But exactly what happened next is unclear. It is known that the first round of experiments centred on inseminating female chimpanzees with human sperm. However the methodology involved has been the subject of lurid speculation. Ivanov always refused to discuss his methods beyond stating that the donors had all been African men who had been 'well paid for their input' – an unfortunate term given the nature of all contemporary and subsequent speculation.

The second round of experiments centred on African women inseminated with sperm donations from chimpanzees. But, again, just how many times this was tried and failed is unclear. By 1927 the French were disturbed by the unsettling rumours that had begun to circulate and they obliged the Ivanovs to return to Russia, where they were given facilities in Sukhumi in Stalin's native Georgia. However, none of their experiments bore any results and, in the time-honoured tradition of Stalin's regime, the elder Ivanov was rewarded with a one-way ticket to exile, in which he died a couple of years later.

DEVASTATING CONSEQUENCES

But the fall-out from his folly may still be with us. Although HIV and AIDS was, until recently, thought by most to be a disease of the 1980s, it is now clear that SIV, the simian equivalent, somehow jumped the species barrier to become HIV and did so some time in the late 1920s, with Guinea one of the strong favourites for the location.

When HIV was first identified as a variant of SIV, rumours abound that the barrier had been breached by humano-monkey sexual activity in the jungle; more conservative medical opinion suggested careless handling of bush-meat kills by hunters was responsible. But hunters in Africa have

HALF-MAN–HALF-APE ARMY

Although there is no evidence that Ivanov knew of the Kremlin's motive for funding his bizarre set of experiments, it has since come to light that Stalin and Lavrentiy Beria, his secret police chief, were keen to see if it would be possible to breed a whole new half-man–half-ape army. When not labouring uncomplainingly to harvest the country's resources in the inhospitable north, they would make ideal soldiers, because chimps are instinctively tribal and much given to organized war in their natural environment. This, coupled with their ability to out-run Olympic sprinters and single-handedly pull up to 1000lb on a torque bar, would have made Humanzees a fearsomely brutish and perhaps mindlessly obedient force. At least, that was what Stalin and Beria had hoped for.

been hunting and butchering chimpanzees and other Old World monkeys for thousands of years, so why would it have taken so long for SIV to make the jump?

In the recent past, medical opinion has been reappraised and is looking again at the HIV implications of the experiments of Ivanov and those of his equally misguided Russian contemporary, Serge Voronoff (see Hey, Hey, We're the Monkeys on page 52).

BACK TO THE BEGINNING

One final word on the descent of man: for some time now, Africa has been regarded as the cradle of modern humanity. This notion moved from academic circles and into the public consciousness in 2001 with the publication of Bryan Sykes' intriguing work of non-fiction *The Seven Daughters of Eve*. Sykes estimated that the first real human beings emerged in Central and Eastern Africa approximately 200,000 years ago.

But anthropologists will keep digging. In 2006 human teeth and countless animal bones bearing distinctive marks that indicated flesh had been stripped from them with flint implements were found in the Qesem Caves at Rosh Ha'Ayin in Israel. The remains were dated 200,000 years earlier than any African find. So, unless Africa can come up with an earlier bid, the Holy Land holds the chalice – for the time being, at least.

Serving One's Fellow Man

African and Polynesian societies indulged in cannibalism

THE CONCEPT OF cannibalism, that man could commit perhaps the ultimate crime by eating a fellow human being, has fascinated civilizations since ancient times. Hindu and Greek mythologies teem with examples of gods eating their own children, or vengeful wives serving up the favourite son in a stew to the unwitting and doting father. Today, people still tell the focus of their desires that they could eat them, and there are numerous other edible metaphors for sex.

PAPAL TYRANNY

There was also in ancient times the well-established trend that required the devoted to sacrifice and devour a surrogate for whichever god or demi-god they worshipped. However, these would have been individual acts of anthrophagy (the eating of human flesh) conducted in a respectful and almost reverential way: the ingestion of blood and selected body parts, such as the heart, only. Usually this was done to ensure the passage to heaven, or even rebirth, of the consumed. Although such concepts now seem primitive, today's Christians routinely partake in metaphorical anthrophagy: the wine-and-wafer ceremony of the Mass or Communion is thought to represent 'the blood and body of Christ'.

While there is a well-established history of such individual acts in ancient times, there has also existed a long-standing belief in the existence of countless African and Polynesian societies that indulged in routine anthrophagy as a part of their staple diet. Not challenged until the late twentieth century, this broadly accepted anthropological 'truth' formed the basis of many an expedition, followed by a highly lucrative lecture tour that brought the concept into the public awareness. Even today, no film of the *Indiana Jones* or *Lost World* genre is complete without at least one scene of someone being popped into a pot on a low heat with suitably gruesome and bone-bedecked savages dancing round their dinner-to-be. And who is responsible for the spread of such ideas? The fifteenth-century Vatican, and its avaricious rush to lay claim to the rest of the world.

THE AGE OF EXPLOITATION

During the Age of Exploration, from the fifteenth to the seventeenth centuries, the Catholic countries of Spain and Portugal stood at the helm of the discovery of the New World. In addition to the gold and silver mined in these new lands, the profits to be had from slavery were enormous. Understandably, the Vatican felt in need of a veneer of justification for this slave trade, which itself presented a problem. Numerous popes and cardinals owned slaves, and the galleys of the papal navy ran on chained labour. But although the Bible abounded with divine justification of slavery of every kind – including the sexual slavery of girls as young as ten in Numbers: 31, where Moses instructs the Israelites to 'keep alive for yourselves' all female children 'that hath not known a man by lying with him' – the Vatican was not anxious to make this known to the populations of Europe at large.

An illustration of the landing of Columbus

It must be remembered that this was a time when the Good Book was available only in Latin and any non-cleric who so much as peeked inside its covers would be burnt alive. Anthrophagy, it seemed, presented the ideal solution. No indigenous peoples of new lands could be enslaved – we are all God's children, after all – unless they were found to be man-eaters, in which case they were not God's children but to be held lower than the beasts of the fields over which the Lord had given man dominion. Therefore man must have dominion over beings that ranked below beasts.

CANNIBALISM IS BORN

Before unleashing the rapacious Columbus on the unsuspecting peoples of foreign lands, Spain's Queen Isabella I of Castile and her husband Ferdinand II of Aragon sought clarification on the matter from Pope Alexander VI, also known as Roderic Borja, a man not unaccustomed to employing a little brute force (see Plane Stupidity on page 75). He was only too pleased to clarify, constructing the notorious Demarcation Line down the map of the world, with the Spanish to be kept to the right of it, and Portuguese to the left.

In 1492 Columbus duly headed for the Americas. Upon his arrival in the Caribbean Columbus cynically pronounced the indigenous Canniba people to be wanton man-eaters and dilligently set about their slaughter or enslavement. As this was the first major exercise of such a nature, the unfortunate natives' name gave rise to the term cannibal.

An artistic representation of Columbus' landing from a book dated 1891

Naturally, as soon as the explorers made landfall on a new country, they pronounced the entire population to be man-eaters – and broke out the manacles, or worse. On Haiti alone, the Spanish reduced the indigenous Taino population from 500,000 to 350 people in a mere thirty years.

To justify their preposterous claims, the enslaving expeditions published lurid pamphlets on their return, complete with etchings of trussed captives looking glum in their cooking pots. These pamphlets would become standard reference material for the early anthropologists of the eighteenth and nineteenth centuries, who set out in search

An early representation of South American natives

of suitably terrifying tribes. But their quests never quite came to fruition – no single anthropologist ever visited a village in which the occupants acknowledged themselves to be cannibals.

THINGS DON'T ADD UP

It is extraordinary that these early anthropologists failed to ask basic questions that would have revealed the truth behind the fiction. How, for example, did Stone Age African or Polynesian tribesmen possess iron pots big enough to cook up the meat harvested from three or four victims? They also failed to work out the maths: these were people who led an extremely harsh and active life without the comfort of an abundance of food. The average lithe native would produce up to 10lb of meat at best, which would leave a village of one hundred adults requiring about ten victims for one decent meal. In the space of one year, the occupants of that one cannibalistic village would eat their way through something in the region of 4,000 of their neighbours – hardly a realistic prospect, given the population levels of such areas.

Anthropologists also failed to address the issues of practicality: the animals of the forests and the fishes of the rivers would have contributed a much higher meat yield, and they would not have shot back. Medical implications involved in eating one's fellow man would also have taken their toll. Eating humans is not good for you. Humans carry a dangerous prion, a protein particle thought to cause brain disease, similar to that which causes CJD or 'mad cow disease'. The Fore tribe of Papua New Guinea, one of the few tribes known to conduct funeral cannibalistic rites, nearly wiped itself out in the late 1950s with a fatal condition similar to CJD, known locally as *kuru*, or the shaking death.

Anthropologists have certainly found ancient human bones bearing marks they claim were caused by butchery; but they cannot know the circumstances in which these injuries occurred. Starvation cannibalism is far from unknown in the apparently civilized West. The first English settlers of Jamestown, Virginia resorted to eating each other when food supplies ran out in 1609; the Russians dined on one another in the Siege of Leningrad during the Second World War; and the diet of those who survived the Andean plane crash of 1972 is now common knowledge. Indeed, the vast majority of authenticated acts of cannibalism have involved white Europeans.

A PECKISH PACKER

The first prosecution for starvation cannibalism occurred in nineteenth-century America with the celebrated case of Alfred 'Alferd' Packer (1842–1907). In November 1873 Packer foolishly led a party of gold prospectors from Gunnison, Colorado, up into the high country where, predictably, the weather trapped them in their shack. Packer soon realized that the only food he was going to find before the thaw set in would have to be provided, quite literally, by his clients. When Packer returned to Gunnison looking surprisingly well fed considering his ordeal, questions were asked ... and the answers turned everyone pale.

Packer's subsequent trial is famous in America, not least for the presiding judge's summing-up. Judge Melville B. Gerry, a Democratic in the predominantly Republican state of Colorado, clearly felt a personal grudge against the defendant. Ordering Packer to stand for sentencing, Gerry said, 'Damn you, Packer, there were only seven Democrats in the whole of Hinsdale County and you, you low-down son-of-a-bitch, ate five of them!'

Rats Get a Bad Press

The plagues of the Middle Ages were bubonic and carried by rats' fleas

Throughout the Middle Ages a vast swathe of plagues ravaged much of Europe. Collectively labelled 'The Black Death' – a phrase not coined until the early nineteenth century – the last of these pandemics struck in 1665, immediately prior to the Great Fire of London. Long-standing opinion has suggested the plague was bubonic – transmitted by the rat flea and infecting the lymphatic system, bringing the victim out in painful buboes (swelling of the lymph nodes). But rats, it seems, have received a bad press for too long.

SMELLING A RAT

The role of the rat flea in the transmission of the bubonic plague was not recognized until 1894. According to Dr James Wood, an anthropologist and demographer who specializes in the spread of epidemics, the bubonic plague itself cannot be reliably traced to any time before the late

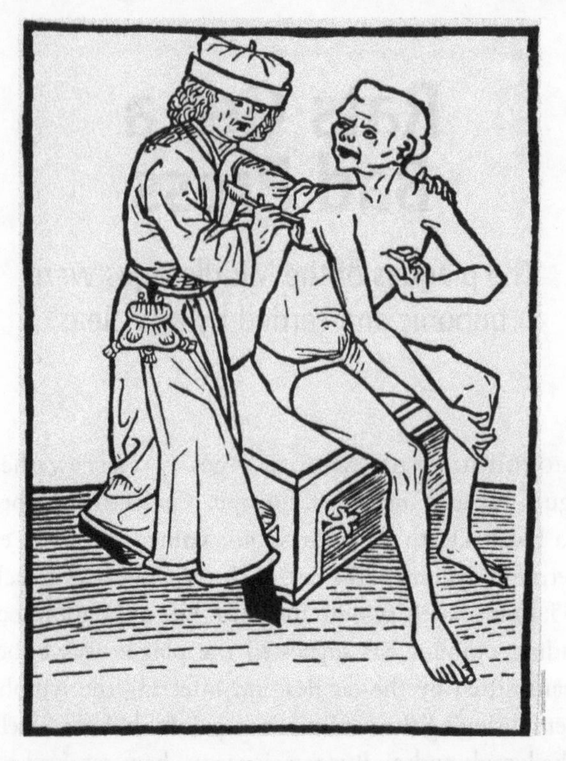

A doctor lances the buboe of a bubonic plague sufferer

eighteenth century; doubtless it existed before that point but no one yet knows just how far back it goes.

Nevertheless, the bubonic theory held sway until well into the twentieth century, when a few academics and epidemiologists began to question this presumption. This group theorized that the plague was pneumonic, an anthrax-like infection that targeted the respiratory system and was transmitted from person to person. While the opinions of individuals in this group differed as to the exact nature of the plague, they were united in the one basic premise: the only theory that did *not* hold water was the established one that blamed bubonic infection.

GREAT BALLS OF FIRE

The plague is to thank for so few dying in the Great Fire of London in 1666 – as few as six deaths were verified. When the plague arrived in the metropolis, the population of the city stood at approximately 650,000, half of whom immediately decamped to other parts of the country, taking the infection with them. The plague wiped out roughly half of those who remained in the capital, so by the time the fire broke out, London was a virtual ghost town.

From the seventeenth century until recent times, mainstream opinion has hailed the fire as the necessary purgative that burnt the plague out of the city. However, the overcrowded slums that were the plague's stronghold were largely untouched by the fire, and the pestilence died out in all other major European conurbations without any conflagration. In addition, the fact that fleeing Londoners took the sickness with them contradicts the long-standing belief that the plague was bubonic.

RACING AHEAD

There are many other problems with the bubonic-based rat-flea premise. The alarming speed at which the plague raced

A victim of the plague is flanked by helpers

across Europe suggests rat fleas did not aid its transmission. Not only did the plague move on a broad front at the rate of perhaps ten to fifteen miles per week, it also threw 'spikes', erupting in sporadic locations many miles ahead. The plague remains a modern affliction, still attacking areas in the tropics and the subtropics in particular. Yet despite modern transport and the increased mobility of people today, more recent outbreaks of bubonic plague have advanced at no more than fifteen or twenty miles per year, killing a mere 3 per cent (with treatment) of the afflicted. One such outbreak occurred in October 2004 in the Indian village of Dangud in

Uttaranchal. The plague was quickly contained and resulted in just three casualties out of a total population of 332; the infection didn't spread to the surrounding villages, even before medical help arrived. How could this possibly be the same beast that had at one time moved a hundred times faster and killed 40 per cent of those in its path?

The puzzling immunity that arose in resistance to the plagues of the Middle Ages also fails to tally with modern-day outbreaks. At times during the Middle Ages the plague struck year after year, a not uncommon affliction. In these instances the mortality rate would fall from 40 per cent to sometimes just 4 per cent in a matter of two generations, eventually only attacking children who had yet to build up their immune systems. However there is no evidence to suggest modern-day victims show signs of developing immunity after surviving their first encounter with recent outbreaks of the bubonic plague.

Finally there is the question raised by the rats and the fleas themselves: why was the plague just as widespread and just as deadly in countries like Iceland, which rats were not known to inhabit until the nineteenth century? And, most puzzling of all, why was the plague just as active in climates outside the temperature and humidity tolerance of fleas, which can only exist and breed within certain parameters?

UNDER QUESTION

Since the 1980s the anti-bubonic lobby has grown increasingly vocal and pointed in its questioning, much to the ire of those who still hold the bubonic theory dear.

The first significant alternative thinking came from the well-respected British zoologist Dr Graham Twigg who, in his book *The Black Death: A Biological Reappraisal* (1984), considers anthrax to have been the culprit. Others, including the medical statisticians Susan Scott and Christopher J. Duncan of the University of Liverpool, followed with *The Biology of Plagues* (2005), in which they consider the plague to have been an Ebola-like virus that was passed on from human to human, while the bubonic plague is almost exclusively passed from the rat flea to humans on an individual basis. Other theories, including the involvement of a type of pulmonary anthrax, are also explored in *The Black Death Transformed* (2002) by Samuel Kline Cohn, Professor of Medieval History at Glasgow University.

Despite the overwhelming evidence to the contrary, those who have dared to question the bubonic theory have been dismissed as 'plague deniers'. However, no 'plague denier' has ever denied that the Black Death turned up with monotonous regularity – they have only questioned the nature of the contagion.

JUST IN CASE …

The best thing a city can do if the bubonic plague comes to visit is to import as many rats as possible and turn them loose in order to give the fleas somewhere to go. This may sound silly, but it works.

Let us assume that the agent was indeed bubonic and that the vector was the rat flea. Rat fleas only move from their chosen host after they have infected and killed it. When the rat population is seriously depleted by such means, the rat fleas move to their next option – cats and dogs; only when these too drop in number do the fleas resort to humans. Naturally, this would mean a sizeable depletion in the populations of rats, cats and dogs prior to any incident of bubonic plague in humans. Yet no chronicler, from the fourteenth century to the seventeenth century, has mentioned whether towns or cities found themselves knee-deep in dead rats, cats and dogs prior to an outbreak.

THEN AND NOW

Let us compare the known symptoms and manifestations of bubonic plague noted during attacks of the Black Death. Modern bubonic infections present buboes in the groin only. Fleas rarely bite humans any higher up the body than the ankle and the nearest lymph nodes to receive the infection are in the groin. Medieval chroniclers talk of buboes appearing all over the victims' bodies, even behind the ears – something that never happens with a bubonic infection.

Medieval medics also recorded weeping sores and abscesses, atypical of bubonic infection, and black pustules, again atypical of bubonic infection but quite typical of anthrax (hence it is derived from the Latin for coal and is thus a sister-word of 'anthracite').

WELL I NEVER! POPULAR SCIENTIFIC IDEAS DEBUNKED

- Lemmings do not hurl themselves off cliffs.
- A bisected worm does not become two viable creatures.
- Cheetahs cannot run at 70mph.
- Hyenas are not dogs but cats.
- Boa constrictors do not crush their prey – they suffocate them.

A doctor's plague habit

Now we should consider the speed of the pandemic spread. Rats tend to nest once they have found somewhere suitable to inhabit, only moving on a matter of perhaps a mile or so when conditions are disrupted or the food supply dries up. This could explain the slow progress of modern outbreaks of bubonic plague, if spread by rats, but hardly fits with the speed of progress of the Black Death. And, if rat-carried, how did the Black Death cross the Alps, the Pyrenees, and make the long journey to rat-free Iceland and Greenland? The answer could perhaps be on a ship, but the length of time such voyages took, coupled with the known incubation period of the bubonic plague, would suggest records would describe ships full of dead rats – and equally dead sailors – washing up along the shores of Europe. Yet no such accounts are in evidence.

Also, if rat-carried, why did the Black Death always fan out along trade routes, and why did quarantine to prevent the spread of infection seem to work? Surely, if the vector was the rat then quarantining any already-infected victim would have had no effect at all, since the infection is so rarely passed from one person to another. That being the case, why was the mortality among doctors and clerics tending to the sick so extraordinarily high?

And what of the rat flea itself? They flourish in high humidity and at temperatures between 50 degrees Fahrenheit and 78 degrees Fahrenheit. Yet when Professor Cohn examined all outbreaks of the Black Death and cross-referenced them to local climate and conditions, they largely coincided with conditions so hostile to the rat flea that numbers would have been at their nadir or non-existent.

A CASE IN POINT

The ever-dwindling bubonic lobby invariably points to the case of the village of Eyam in Derbyshire as proof of their theory. Towards the end of August 1665, local tailor George Vicars received a delivery of cloth from London. On opening the parcel he supposedly released rat fleas infected with the disease, which led to his death on 7 September. However, those who favour the anthrax theory are quick to point out, quite rightly, that the spores can remain active in infected wool, cloth and other animal products for years.

Whatever the nature of the Black Death that came to Eyam, Vicars had already passed it on to several others before dying, and again it has to be stressed that this is not possible with bubonic plague. The tradition goes that the doughty

EYAM MYTHS

And what of the use of the vinegar dip, which also crops up in similar tales? People simply did not know about germs, bacteria and viruses in such times; it was not until Louis Pasteur's experiments of the 1870s that medical circles accepted the germ theory (see Heaven Scent on page 106). Prior to that time sickness was thought to be caused by bad air and bad smells. Why, therefore, were the Eyamites two centuries ahead of medical knowledge with their vinegar dip?

The Eyam Black Death legend is also responsible for the oft-heard myth that the rhyme 'Ring-a-Ring o' Roses' was created by village children watching the deadly plague drama unfold. However, no contemporary records mention red blotches or violent sneezing and, according to the *Oxford Dictionary of Nursery Rhymes*, the rhyme was unknown before the 1880s, when it was imported from New England.

villagers, under the guidance of their rector, William Mompesson (1639–1709), observed a protracted and self-imposed quarantine to halt the progress of the plague and prevent it spreading to the surrounding villages. According to the tradition, all stayed put to take their chances, with food and other necessities being brought to pick-up points by people from the villages that the Eyamites were trying to save.

Not wishing to take their gratitude to ridiculous lengths, these other villagers demanded payment for the goods they delivered, and present-day guides at Eyam will solemnly point to depressions in stones at the exchange points, pronouncing these to be the much-fabled 'vinegar dips' to avoid the coins passing on the infection. The quarantine lasted fourteen months, after which only eighty-three of the original 350 remained alive. Had the visiting infection really been bubonic, then quarantine would have been absolutely useless, unless it had been imposed upon the rats.

It is perhaps unfair to highlight the Eyam case because there is no contemporary documentation to back up the tale. A captive population of 350 would not long hold the attention of a single, fast-killing infection – certainly not for fourteen months. It is also known that several local families did indeed flee the area, with Mompesson himself packing his two children off to Sheffield.

Who's Your Daddy?

Offspring can inherit characteristics of previous mates of the female parent

HUMANS ARE PROGRAMMED to procreate, but some people take more time than others in finding that special someone with whom they want to start a family. The theory of telegony would suggest you should choose your partners wisely, for this particularly spurious branch of scientific theory believed offspring could inherit characteristics of the mother's previous sexual partners.

THE IDEA IS BORN

Dating back as far as Aristotle, and not discredited until the late nineteenth century, telegony was born of the notion that the female served only as a vessel in all matters sexual and procreative. It was believed that woman received a permanent 'imprint' from all of her sexual partners; the imprint from the first partner was deemed to be the strongest, and all subsequent partners imprinted her in descending order.

While the theory might sit well within misogynistic circles, telegony may have once helped to demystify atypical pregnancies, the causes of which were not understood until the invention of DNA profiling in 1984. Thanks to advances in science, it is now known that, in particular circumstances, a woman can deliver a double birth in which the apparent twins have two separate biological fathers. The most recent case was that of Mia Washington of Dallas, Texas, who in May 2009 discovered, through a routine medical procedure, that her eleven-year-old twins were of different paternity.

It transpired that at the time of conceiving her children Mia had been having an affair, two-timing her partner with another man and having sex with both men on a regular basis. Because sperm can remain alive inside a woman for up to five days, the result was that both men had fertilized an egg within perhaps hours of each other. Because all the people concerned were of black origin, no alarm bells were raised at the time of birth; it was not until a routine health check eleven years later that the truth was finally revealed.

There have been cases when both a black child and a white child have been born to the same mother at the same time. One such case involved a white German woman in post-war Berlin who was sleeping simultaneously with her white German lover and a black American GI. One can only imagine how the ancients might have perceived such a thing should it have occurred two thousand years ago. Although whether or not such events were the inspiration behind the theory of telegony will never be known.

TAILOR-MADE TELEGONY

Different societies interpreted the rules of telegony in different ways. Ancient Jewish tradition permitted a complex type of marriage that resulted in the biblical character Onan (whose reputation gave rise to the solitary sexual act known as onanism, or masturbation, see box opposite). The little-understood aspect of Torah Law, known as a 'Yibbum marriage', stated that should a woman be widowed while still of child-bearing age, a brother of the departed was obliged to marry her, whether he was already married or not. His role was to raise sons in order to qualify for the deceased brother's portion of the family inheritance. Jewish custom deduced that, within the confines of telegony, the widow was already imprinted by the deceased, so who better than a telegonic replica, such as a brother, to continue the process.

In medieval England the theory of telegony held strong and formed the foundation of the main objections to the marriage of Edward, The Black Prince, to Joan, 'The Fair Maid of Kent'. Telegony dictated the value of a virgin bride who came 'unprinted' and, having first been married at the age of twelve and again at the age of twenty, Joan was already sexually experienced. Should a royal or noble have chosen to ignore the curse of telegonic contamination of his children and insist on marrying a widow, a divorcee or a commoner, then his only option was a ritual called a morganatic marriage.

This convenient side-step required the groom to give the bride a single gift first thing in the morning (or *morgan*, as in

TELEGONY IN THE BIBLE

In the Bible, Onan was required to marry his sister-in-law, Tamar, after God had killed her husband. Onan was a rather greedy chap and he failed to see why he should produce heirs for his dead brother's inheritance share, which would otherwise fall to him. Far from indulging in solitary pleasure, as legend has it, Onan did have intercourse with Tamar but withdrew at the last minute to 'spill his seed upon the ground' (Gen. 38:8–10) and thus avoid inseminating her. God killed Onan too, so his father, Judah, thought he should carry out the job himself, which, of course, sat well within telegonic theory.

the early German word) after the wedding night – that being the only benefit she could expect from the union. Most importantly, any child born of the marriage was excluded from the lineage of the father's family and regarded in law as but one step above illegitimate.

Morganatic marriages proved popular among a handful of notable royals. The Austrian Archduke Franz Ferdinand, heir to the Austro-Hungarian throne, whose assassination at Sarajevo in 1914 led to the outbreak of the First World War, was forced to undergo such a marriage to his beloved Sophie Chotek. Because Sophie was not the member of a reigning or formerly reigning dynasty of Europe she was not deemed an eligible partner for the Archduke. But he was deeply in love and his persistence paid off when he was eventually allowed to marry her after Emperor Franz Joseph of Austria permitted a morganatic marriage. The British King Edward VIII's reign was cut very short after he tried a similar venture in his attempt to wed American divorcee Mrs Wallis Simpson; blocked by Prime Minister Stanley Baldwin, Edward decided to abdicate.

THE IDEA GAINS GROUND

The English philosopher and biologist Herbert Spencer, the man responsible for coining the term 'survival of the fittest' (see Mendel to Mengele on page 57), thought telegony a valid theory, as did the philosopher Arthur Schopenhauer (1788–1860). However, Spencer also thought phrenology to be a valid science, and Schopenhauer's ideas of women in general – his 1851 essay 'Of Women' commented that 'woman is not intended for great mental or for great physical labour' – were equally questionable. Even Darwin himself fell into the telegony trap after the famous case of Lord Morton's Mare, as it came to be known.

In 1821 George Douglas, the sixteenth Earl of Morton, reported to the Royal Society, one of the world's oldest scientific academies, that after he had crossed a quagga (a now extinct type of zebra) with a mare something rather strange had happened. When the same mare was later bred with a white stallion, the foals born had striped markings on the legs, just like the quagga with whom the mare had previously procreated. While it was most likely to have been

An accidental hybrid: an illustration of a quagga

a coincidence – and I can find no record of any foal being independently examined – Morton's word was taken as proof of telegony, such was his standing. Darwin certainly believed the story and mentioned the alleged incident in his *On the Origin of the Species* (1859) and in other writings.

Naturally, racist elements were quick to cite telegony as the scientific justification for encouraging white girls to avoid contact with 'undesirable elements'. From its hideous inception in post-Civil War America up to the 1950s, the Ku Klux Klan's message to the girls of the American South was that just one kiss from a black man would contaminate their fertility and render them likely to bear a black baby, even if they were married to a white man.

The rediscovery in 1900 of Gregor Mendel's unprecedented research into the field of genetics (see From Mendel to

Mengele on page 60) signalled the death knell for telegony. The father of modern genetics, Mendel's experiments proved how traits are inherited across the generations. These laws of heredity exposed the folly of telegony and banished it to the annals of science. Or did it?

Despite the rediscovery of Mendel's work, telegony still had, and continues to have, its supporters. Much like the Ku Klux Klan before him, Hitler made use of the theory to frighten girls away from contact of any kind with non-Aryans. As recently as 2004, the Russian Orthodox Church published *Virginity and Telegony*, which cautions young girls to keep themselves 'intact' in order to save themselves from past sins haunting them in the form of babies delivered in marriage but bearing resemblance to a past partner. The imprint of telegony still persists within the world of dog-breeding, where the concept that a pedigree bitch, once 'contaminated' by a litter of pups from another sort of dog, can never again produce examples of her own breed worthy of showing still prevails.

Subterranean Homesick Blues

The earth is a hollow vessel

Early religion touted the notion of a hollow earth; a vessel that housed the 'land of final punishment' for anyone that failed to conduct themselves in accordance with an individual religion's various dictates. For the Ancient Greeks hollow earth was the Underworld, ruled by the god Hades, where all people ended up, good or bad. For the Christians it was, and still is, the place we call Hell.

SCIENCE GETS INVOLVED

Inspired by theological pronouncements of an underground world, early science took the theory to heart and further explored the concept. According to the proponents of the hollow earth theory, our planet possessed a variety of characteristics, including an interior sun at its core, a living population, and access points at the Poles, or in Tibet. Opinions on the exact details of the hollow earth's make-up differed according to different sources.

TURNING UP THE HEAT

Mark Twain once remarked, 'Heaven for the climate, but Hell for the company' but, according to the Bible, Heaven is the hotter of the two. Isaiah 30:26 describes that in Heaven 'the light of the sun shall be sevenfold as the light of seven days' which, according to a series of weighty calculations involving the Stefan-Boltzmann fourth power law of radiation, gives us a temperature of 525 degrees Celsius. Hell, the Bible tells us, abounds with pits of liquid brimstone (sulphur), so the temperature there must be 445 degrees Celsius, because any temperature higher than that turns sulphur into gas.

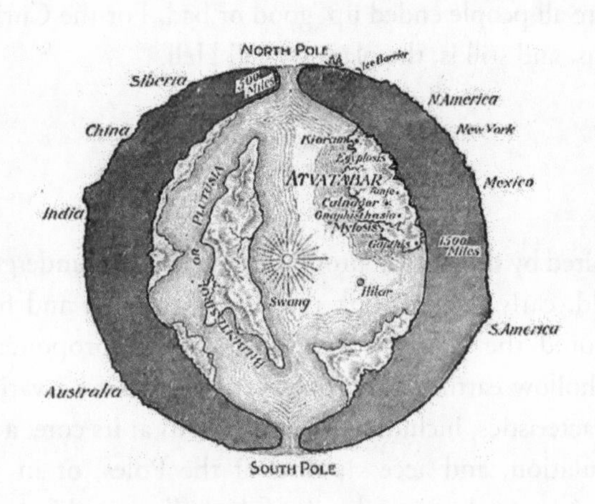

The hollow earth with access points at both Poles

In 1692 Edmond Halley (1656–1742), the astronomer and polymath of comet fame, advanced a theory that suggested the planet was a hollow shell that contained three inner shells. He believed each independent and inhabited sphere was separated from its neighbouring sphere by its own atmosphere. The access points to the interior of the earth were, according to Halley, located at the Poles, and gases escaping from within the earth manifested themselves as the aurora borealis. Halley's hypothesis drew considerable recognition.

It was Halley's 1676 voyage to St Helena to observe the star constellations in the southern hemisphere that started him thinking seriously about the ancient hollow earth theory. He was struck by the inconsistencies in his compass readings, which seemed to vary even on the same spot on consecutive days. What else could account for this except another rotating sphere, or two, within the casing of the planet? Of course, what Halley did not then know was that this discrepancy in compass readings is quite normal: the lines of the earth's magnetic field do not run in straight lines between the Poles, but in a series of erratic lines, which themselves shift.

The old-fashioned needle compass aligns itself with whichever line is nearest, and thus never points to true North, unless by pure coincidence. The lines are generated by the rotating, solid iron ball at the centre of the planet, which is, at present, behaving in a rather worrying way. Things, it seems, are afoot beneath our feet and some feel the earth is preparing for a reversal of its own polarization. This change happens approximately every 250,000 years, and we are long overdue such an upheaval.

The idea that the lines of the earth's magnetic field were caused by something deep within the earth was the thrust of the first part of Halley's 1692 submission to the Royal Society. If one allowed for the notion of other spheres that possessed their own magnetic fields, which rotate at different speeds or even in different directions, then this would account for the varying compass readings. Halley informed the assembled members of the Royal Society that his notions fitted with the theological perspective that the Almighty would not, in His wisdom, have created the massive bulk of the world 'simply to support its surface' but instead 'to yield as great a surface as possible for the use of living creatures as can consist with the conveniency and security of the whole'. When asked why the oceans did not drain away into the hollow after sub-oceanic earthquakes tore open cracks in the earth's surface, Halley's answer was that the outer sphere, perhaps 500 miles thick, was obviously self-healing due to the presence of 'saline and vitriolic particles as may contribute to petrification'.

Halley concluded his submission by saying that 'The concave arches may in several places shine with such a substance as invests the surface of the Sun and I have adventured to make these subterranean orbs capable of being inhabited.' From those last few words on the subject there sprang a whole new interest in hollow earthism.

DEARLY DEVOTED

The list of those who took Halley at his tantalizing word is long to say the least, but among the more prominent were the Norwegian-born essayist and philosopher Ludvig Holberg (1684–1754), who enshrined his ideas in the novel *The Journey of Niels Klim to the World Underground* (1741). In a yarn that would long pre-date Jules Verne's altogether more enjoyable *Journey to the Centre of the Earth* (1863), Holberg's fantasy tells the story of a young student who discovers a series of worlds within our own after falling into a cave. His exciting journey leads him to explore new lands and strange creatures that lived under the earth's crust.

The eighteenth-century hollow-earthers also counted among their ranks the Scottish mathematician and physicist Sir John Leslie (1776–1832). In his *Elements of Natural Philosophy* (1829) he devotes nearly half a dozen pages to the concept. In the nineteenth century the leading lights of the lobby were the Americans John Cleves Symmes (1779–1829), James McBride (1788–1859) and Jeremiah Reynolds (1799–1858). The first is best described as a well-connected adventurer, McBride was a leading light of Miami University, and Reynolds a well-respected newspaper editor and explorer.

Together the three men lobbied President John Quincy Adams (1767–1848) who, as a proponent of the hollow earth theory, agreed to fund from the public purse an expedition to the South Pole to find the portal to the supposed subterranean wonders. However, the project was abandoned when Adams' replacement, the altogether more hard-headed Andrew Jackson (1767–1845), threw the idea and its proponents out on their collective ear.

Undeterred, Reynolds immediately raised funds for the expedition from private backers and speculators, and set sail in late 1829 in search of his goal. His crew, however, were unconvinced and, having quickly tired of looking for a big hole in the earth's surface under the supervision of a man they thought unhinged, they mutinied and set Reynolds ashore on the coast of Chile, before sailing off in his ship. Reynolds was forced to wait in Valparaiso until 1832, when he was eventually rescued by an American ship.

NAZI INVOLVEMENT

Not one to let a dubious scientific theory pass by unnoticed, Adolf Hitler too propagated the theory of a hollow earth. The Thule Society, a large collection of right-wing eccentrics who adhered to all manner of mystical notions, provided the hub of support in Germany. It was from this semi-clandestine society that the Nazi Party itself sprang. The Thulians believed the access to the inner world lay in Tibet, a country they also believed to be the cradle of a long lost master race; Hitler and a majority of his henchmen also believed this to be true.

In 1938 Hitler and Himmler sent an expedition to Tibet to find anthropological evidence of the supposed master race and, while they were there, to investigate any suspiciously large holes they might happen to encounter. The expedition heard much talk of the fabled underworld cities of Agartha and Shambhala (now known as 'Shangri-La' to Westerners) and the superior beings who lived there. Despite the explorers leaving their first expedition empty-handed, Hitler had not yet finished with the Underworld. In 1943 the Führer decided it was time he explored the other hollow

JESUS LIVES

Both the Thule Society and the Nazi Party explored an obsession with symbolism and the quasi-occult. Hitler too was a proponent of such beliefs, dispatching minions in search of the Holy Grail, the Ark of the Covenant and the so-called Spear of Destiny, which, according to John's account of the crucifixion, was a lance used to pierce Jesus when he was tied to the cross in order to confirm his death. Unfortunately the author of the Gospel of John – written sometime in *c.*100AD, and thus far too late to have been an eyewitness account – was caught off guard by outdated medical knowledge.

The Ancient civilizations believed the arteries were responsible for the movement of air around the body (hence 'a(i)rtery'). Their mistake was no doubt based on the fact that early anatomists always found the arterial system to be empty and therefore failed to grasp the fact that, once the heart stops generating pressure and a person dies, blood retreats to the veins. John 19:34 states that from the spear, 'came there out blood and water' but, as all fans of popular television forensic series know, corpses do not bleed, no matter how many times they might be lanced. So, far from validating the story of Jesus' death on the cross, it confirms the reverse: Jesus was alive at the end of the crucifixion, therefore there could have been no resurrection.

earth theory: the earth is a concave sphere and all living creatures inhabit its inside surface.

In April 1942 Hitler sent an expedition under the leadership of Dr Heinz Fischer to Rügen Island, located in the Baltic Sea, where they set up camp with powerful telescopes and radar. They were under instruction to aim their instruments up at the sky, rather than across the sea, to spy on Allied activity on the other side of the world. Unsurprisingly, the members of the expedition returned to Berlin in late May empty-handed, and no doubt fearful of reprisal from their dogmatic leader. Fortunately for them, the high-ranking Nazi Reinhard Heydrich had only recently been assassinated in Czechoslovakia and Hitler was far more interested in plotting his revenge. Relieved, Fischer took himself off to secure obscurity and all further Nazi-led hollow earth projects were abandoned. But the theory of a hollow earth did not die out with Hitler. There remain today countless hollow earth societies, all of which, of course, believe that the successes of the various NASA-led space programmes are but a charade to keep us from communicating with our inner selves.

Are Bears Polar?

The bodies of animals contain a life energy that can be influenced by external magnetic forces

To suggest someone possesses an animal magnetism is to imply they are sexually attractive or have a charismatic allure, but the original meaning of the expression suggested nothing of the kind. First realized in the eighteenth century, animal magnetism was a new 'science' that suggested people and animals possessed a universal fluid that could be influenced by an external magnetic force. This notion held sway for some considerable time, and would go on to spawn the phenomenon of hypnotism.

ANIMAL MAGNETISM

The Austrian-born student Franz Anton Mesmer (1734–1815) first devised the concept of animal magnetism and he did so under the tutelage of the incongruously named Jesuit astronomer Father Maximilian Hell (1720–92). Aside from his interest in the universe, Maximilian Hell also had a keen interest in the spurious field of magnetic therapy, born in part from his familiarity with the Chinese concept of qi (pronounced 'chee'). This equally bogus notion centred on the ancient oriental belief that certain vital energies flow through the body, and that the disruption of these energies can lead to ill health. Wellness is re-established by redirecting the flow of qi using a variety of techniques, from feng shui to acupuncture and magnet therapy.

The acupuncture needles and magnets were intended to act like corporeal traffic police by redirecting the energy flow to its correct route. Hell had no time for acupuncture; he was convinced the magnets, rather than the needles, held the key to the cure. Hell started to build medicinal magnet theory into his lectures, and his student Mesmer joined him wholeheartedly in his folly.

Mesmer was also greatly influenced by the British Royal Physician Richard Mead (1673–1754), the 'Father of Foundlings' whose spacious residence in Bloomsbury, London, would form the foundation of the Great Ormond Street Hospital for Sick Children. Mead had a keen interest in astronomy and, after his close friend Isaac Newton discovered the universal force of gravity, Mead postulated that just as the planets exerted a gravitational pull on the earth, they similarly influenced the flow of fluids within animals and humans.

While the gravitational pull of the moon creates tides in the earth's oceans, for example, the human body is far too small an object to be subjected to similar effects. Mesmer, however, used the false postulations of Hell and Mead to devise the concept of animal gravitation.

THE INFLUENCE OF THE MOON

Planetary influence on the well-being of humankind was not in itself a new notion: the word 'lunatic', derived from the Latin *luna* (moon), suggested the full moon could induce abnormal behaviour in those who were normally sane. In the days before street lighting there may have been some foundation in such a notion. Taking advantage of the free light provided by the full moon, people might well have stayed out longer, allowing them the opportunity to drink more than usual. Or perhaps they always behaved like that and it was just that their transgressions were more visible to sober observers.

Mesmer next observed the Bavarian priest exorcist Johann Gassner (1727–79) during a routine exorcism. Mesmer concluded that the patient's possession was alleviated by the magnetism that emitted from a metal cross the priest had used to beat and stroke the afflicted. In the belief that there might be an interaction between a natural magnetic quality within the body and the external magnetic force, Mesmer devised the concept of animal magnetism.

Mesmer was convinced a universal fluid, a force that flowed like liquid through all living creatures, could be manipulated by an external magnetic influence. In order to test his hypothesis, Mesmer subjected his patients to a variety of bizarre treatments, one of which included having them sit in a vat of diluted sulphuric acid while they held on to iron bars that carried a low-voltage current.

MESMERISM

By 1775 Mesmer abandoned the use of magnets and electricity altogether, as he had become convinced that he himself was the vector directing and controlling the ebb and flow of a person's universal fluid. By this stage his patients were required to sit in mesmeric cubicles, which, they were told, would concentrate 'the force' on them.

Nor was Mesmer alone in his beliefs – many physicians of significant standing followed his teachings, and 'mesmerism' became popular throughout eighteenth-century Europe. In Mesmer's defence, his research did lead to a groundbreaking discovery; unfortunately he was too enamoured with the notion of animal magnetism to acknowledge the phenomenon for what it really was.

Mesmer enjoyed some considerable success in his sessions, which had become a cross between a pseudo science and spiritual healing. His success, however, may have been the result of his propensity to focus his attentions on hypochondriacs, hysterics and those otherwise susceptible to his suggestions. The most celebrated of such patients was the talented pianist Maria Theresia von Paradis (1759–1824), whose father was prominent in the Austrian court and held the ear of the Empress Maria Theresia, after whom he had named his daughter.

Franz Anton Mesmer at work

Having suffered from 'hysterical blindness' since the age of four, the eighteen-year-old Maria Theresia experienced a temporary improvement in her sight under the guidance of Mesmer. But when it was suggested that Mesmer had used his influence over Maria Theresia to put her into one of his famous mesmeric trances for other, less salubrious purposes, her parents felt they had no option but to dispense with his services. As soon as the treatments were terminated the blindness returned, and remained with Maria Theresia for the rest of her life.

The whiff of scandal was enough to force Mesmer to quit Vienna for Paris, where he continued his lucrative practice. Mesmer began to record that the influence he communicated to his patients induced what he called a magnetic trance, or magnetic somnambulance. Although he did not realize it, Mesmer was hypnotizing his patients.

MESMER IS MARGINALIZED

As Mesmer's bandwagon picked up momentum in France, Louis XVI came under increasing pressure to establish

This illustration adorns the frontispiece of *Confessions of a Magnetizer*, an 1845 exposé of animal magnetism

a scientific inquiry into the matter. Despite being one of Mesmer's patients, Louis XVI finally gave the green light in 1784 for a Commission. One of the members of the committee was Dr Joseph Ignace Guillotin, an expert in pain management whose name would soon be made infamous by the impending French Revolution.

Guillotin and his co-committee members found no merit in Mesmer's methods or treatments; he was denounced as a fraud. Another prominent name on the committee, Benjamin Franklin (1706–90), one of the Founding Fathers of the United States, then in Paris as the United States Minister to France, concurred, although he did add that Mesmer clearly had an idea of the role the mind played in sickness and its cure.

HYPNOTIC

As Mesmer became increasingly marginalized, one of his French students, Jose Faria (1746–1819), an Indo-Portuguese monk from Goa, began to incorporate the techniques used in oriental hypnotism into the sessions. The results were revelatory. Finally realizing that hypnotism, and not animal magnetism, lay at the heart of Mesmer's sessions, Faria pronounced that 'Nothing comes from the magnetizer: everything comes from the subject and takes place in his imagination; [it is] autosuggestion generated from within the mind [of the subject].'

Faria's take on the matter would outlive him. His methods were later observed by the English physician James

Braid (1795–1860) who in 1841 made a close study of this modified form of mesmerism and refined it to become the first modern clinical hypnotism, a term he coined himself.

With this new life breathed into the phenomenon of mesmerism, others sought to apply the concept to the rest of the animal kingdom. The methods employed by North African and Indian snake charmers were re-evaluated – was it just a street trick, or did something more sophisticated occur between the charmer and the charmed? Chickens, too, seemed to fall into a trance if held down against the ground as a chalk line was drawn away from the tip of their beak.

Unfortunately for those who like to believe in such things, animals cannot be mesmerized or hypnotized. The snake-charming acts are nothing but a street trick; the music from the charmer's flute is quite unnecessary to the ruse, as its pitch lies outside of a snake's auditory range. (It is a myth too that snakes are deaf – they simply do not have external ears, only internal ones that pick up vibration through the ground.) The key to the trick is the tapping of the charmer's foot and the movement of his flute.

The strike distance of a cobra is approximately two-thirds of its body length and the creature has motion-sensitive vision. The key to a long career as a snake charmer lies in their ability to position themselves just outside of that range, but not so far that the snake loses interest and slides back down into its basket. The snake appears to mimic the flute's movement from side to side, as if in a trance; in reality it is preparing itself to strike the charmer, should the opportunity present itself.

Chickens, on the other hand, go into a state of thanatosis (feigned death), in response to the sheer fear they experience

A chicken undergoing hypnosis

at being held immobile, as if in the grip of a predator. The carefully drawn chalk line is a mere piece of theatre, quite unnecessary to the procedure.

THE LEGACY LIVES ON

While mesmerism finally waned, and its offspring, hypnotism, went on to achieve greater things, certain charlatans were not yet done with magnet therapy. There still exists today a multi-million-dollar industry that peddles magnetic bangles to the gullible, promising improved blood flow and relief from pain in the wrists and hands. An Internet search for 'magnetic bangles UK' reveals over 2 million pages, and 'magnetic bangles USA' renders nearly 8 million sites.

Nor has Mesmer's original idea of a universal force that flows through the bodies of all animals been completely disregarded. For some, 'orgone', a so-called cosmic energy discovered by the Austrian-born psychiatrist and psychoanalyst William Reich (1897–1957), held the key. For others it was, and still is for some, 'vril', a powerful life-giving and life-destroying force first introduced in Edward Bulwer-Lytton's 1871 novel *The Coming Race*, which some took to be a fictionalized realization of a very real force. Mesmer's ideas even form the foundation of the Star Wars franchise, in which The Force is always with those willing to recognize and harness it.

A close friend of Sigmund Freud and a member of the so-called Vienna Circle, Reich was captivated by Freud's notion of libido. For Reich libido was the universal force, which he chose to call orgone (a blend of 'orgasm' and 'ozone'). Under the pretence he could tap into the vibrations of an enduring cosmic orgasm, Reich built orgone accumulators, largely modelled on the mesmeric cubicles, and into which he invited his human guinea pigs for a quick blast of celestial sex-energy. Reich's accumulator was essentially a Faraday cage (invented in 1836 by the English scientist Michael Faraday for the use in blocking external electric fields), insulated on the outside by wood and lined on the inside with thin steel sheeting.

He believed the orgone accumulator contained concentrated orgone energy which could be used to cure illness in those who were experiencing an imbalance in their orgone.

Participants emerged claiming to feel suitably invigorated; however, Reich did select his volunteers from a circle of the already converted. Interestingly, Reich did attract some notable names to his camp – even Albert Einstein (1879–1955) thought there might be something in it. After a five-hour meeting between the two men at Princeton University in January 1941, Einstein declared that if Reich's experiments revealed a detectable rise in temperature in one of his accumulators without a known heat source, then the basic laws of physics would have been discredited.

Reich returned to Princeton with an accumulator so Einstein could observe for himself the increase in temperature. And observe it he did. Einstein's experiments revealed a noticeable increase in the internal temperature of the accumulator; however, this increase was detectable only in the upper section of the box, and it was soon established the temperature change was the consequence of convection currents from the room itself and not the result of a cosmic influence.

The accumulators were not airtight; the only thing they did accumulate was warm air from without. In 1954 the US Food and Drug Administration (FDA) issued Reich with an injunction, banning him from selling or transporting his accumulators. But Reich was by then a psychotic paranoid with delusions of grandeur and he chose to ignore the writ. He ended his days in a secure unit, a somewhat disgraced figure. Reich's accumulators are still available to buy online for a mere $5,000, which the unkind might say is a touch steep for what is essentially a steel-lined coffin. But many are willing to pay such sums to be at one with the universe.

THE NEXT GENERATION

Bulwer-Lytton's vril was perhaps mesmerism's most sinister spin-off. *The Coming Race* combined the notion of a hollow earth (see Subterranean Homesick Blues on page 155) with the universal force of mesmerism. For Bulwer-Lytton, vril flowed like a magical fluid through the bodies of a subterranean super race that was biding its time for a terrestrial take-over. Such was the international popularity of this yarn it even inspired product names such as Bovril – a blend of 'bovine' and 'vril'.

The Coming Race was a great hit in Germany, especially among certain societies that had started to emerge in the early 1900s. The Thule Society in particular thought the book was fact masquerading as fiction. The Thulians believed the earth was a hollow vessel, within which dwelled the mythical Aryan race, biding its time before bursting forth on to the earth's surface to take over the world.

From the Thulian membership would emerge the Nazi Party (see Subterranean Homesick Blues on page 160). Under the guiding hand of Himmler, in 1935 Hitler set up the Studiengesellschaft für Geistesurgeschichte, Deutsches Ahnenerbe (the Study Society for Spiritual History and German Ancestral Heritage), which was charged with proving the existence of a subterranean super race. Their mission was

WELL I NEVER! POPULAR SCIENTIFIC IDEAS DEBUNKED

- Water does not spiral down the plughole in different directions in the northern and southern hemispheres.
- Goldfish have pretty good memories, and can even be taught a trick or two!
- Black Holes do not absorb surrounding matter.
- Electricity actually flows from negative to positive.
- The Big Bang Theory relates to the early evolution of the universe, not its inception.

to establish contact and reassure the subterranean dwellers that they had allies on the surface who would be there to support them when the time came.

This was the hidden agendum of the infamous 1938 German expedition to Tibet, as led by SS member and German hunter Ernst Schäfer (1910–92) and his Deputy Expedition Leader, fellow SS member Bruno Berger (1911–2009). The official reason given for the expedition was to research the geography and culture of the region, but Berger and his SS colleagues did little more than take cranial measurements of the local people and make phrenological head casts with the results.

The last word on the subject of animal magnetism is awarded to Willy Ley (1906–69), the leading rocket scientist who had the good sense to leave Germany in 1935. In 1947 Ley wrote an article called 'Pseudoscience in Naziland' in which he mentioned a sub-group of the Thule Society that was exclusively founded on the contents of Bulwer-Lytton's *The Coming Race.*

'The next group was literally founded upon a novel,' wrote Ley. 'That group, which I think called itself *Wahrheitsgesellschaft* (the Society for Truth) and which was more or less localized in Berlin, devoted its spare time looking for vril.' Indeed there still exist vril societies throughout Europe and North America today. Perhaps someone should tell them it was all just a story.

Liquid Assets

The body is made up of four humours – blood, phlegm, yellow bile and black bile

FROM ANCIENT GREECE to the mid-to-late nineteenth century the 'humour theory', the idea that the body is made up of four main fluids, was a broadly accepted phenomenon. Consisting

This engraving dates from the sixteenth century and shows the perceived balance of the four humours

of phlegm, yellow bile, black bile and blood, an imbalance in the four humours was said to cause all sickness and mental malaise. Too much phlegm and a person would become phlegmatic; an excess of blood (or *sanguis*) was thought to make a person sanguine; too much yellow bile (or *chole*) would likely induce cholera; and too much black bile (or *melanchole*) and a person would sink into a melancholy mood. These four basic concepts constituted a comprehensive framework of health and disease made available to doctors until it was superseded by scientific medicine in the nineteenth century.

FOUNDATION STONES

The foundation for the humour theory lay in Hippocratic medicine. Beginning with the Ancient Greek medic Hippocrates (*c.*460 BC–*c.*370 BC) and developed by unknown individuals over the course of the next two centuries, the Hippocratic Corpus was later introduced to the West by the physician, surgeon and pivotal figure in Greek medicine, Galen (AD 129–*c.*210). The theory held that each of the four humours was identified with a body part and linked to one of the four elements – phlegm was matched with the brain and water, blood with the heart and air, black bile with the spleen and the earth, and yellow bile with the liver and fire. The properties of the four humours were heat, cold, dryness and moistness.

The Hippocratic approach was holistic – the Ancient Greeks knew little about the human anatomy as they were averse to dissecting human bodies; they favoured instead a surface inspection of their patients to see if they could spot the likely signs of disease. Physicians believed they could read the complex balance of the four humours in an individual's face, hence 'complexion'. The cosmetics trade was in part put into hyper-drive by the humour theory, as both men and women tried to present to the world a visage that spoke well of their physical and mental disposition. As the theory expanded it embraced the notion that you are what you eat. Foods were categorized according to the effect they had on the humours, and this shaped the different cooking styles of medieval Europe (see box below).

FOOD, GLORIOUS FOOD

The different nations of medieval Europe each had a slightly different take on how food might influence a person's humoural balance. Red meat was thought to anger the blood, but if cooked in honey the effect would be reduced; an imbalance in bile could be treated by the chef with foods dressed in saffron. Not only did the humour theory invade the medieval kitchen, the chefs themselves rose to physician-like status, commanding considerable pay and respect.

BLOODY EXCESS

Proponents of the four humours believed the body could heal itself; its procedures were designed to encourage the body in this natural healing process. Bloodletting was a common practice, often carried out to help alleviate fever and to remove excess blood, which continued until well into the mid-nineteenth century. So popular was bloodletting that, no matter their condition, many patients were bled relentlessly by their doctors, often with terminal results. American President George Washington (1731–99), British

There will be blood: a lithograph from 1804

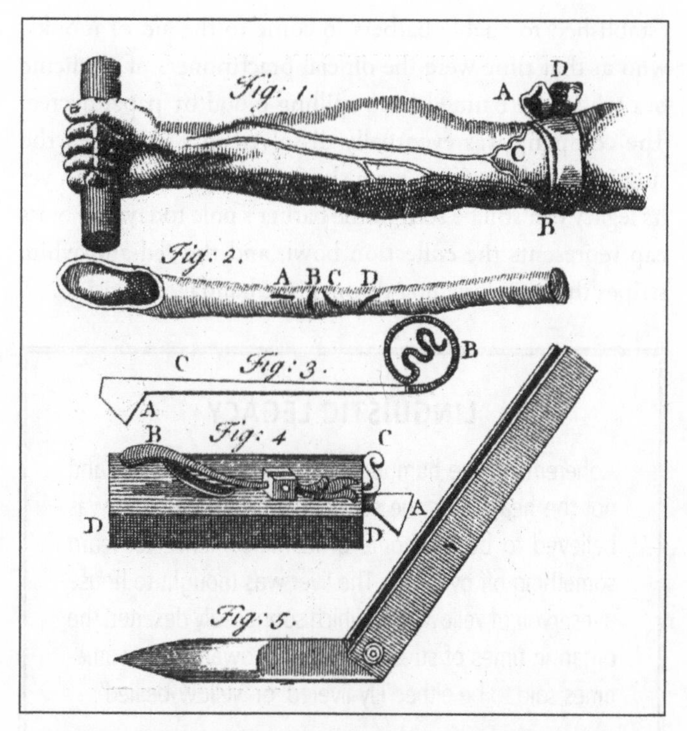

These illustrations of the instruments and techniques used in bloodletting are taken from a 1759 edition of *A General System of Surgery*

poet Lord Byron (1788–1824) and the Scottish poet and playwright Sir Walter Scott (1771–1832) were all bled to death by well-intentioned doctors.

Humoural bleeding intended to improve a person's outlook was so popular that barbers could help patients dispense with a pint or two. In the fourteenth century the Worshipful Company of Barbers and Surgeons was established to enable barbers to come to the aid of monks, who at that time were the official practitioners of medicine but who were banned from spilling blood by papal decree. The company was eventually dissolved in 1745 when the surgeons broke away to form the Company of Surgeons, yet its legacy can still be seen in the barber's pole today: the brass cap represents the collection bowl, and the red-and-white stripes the blood seeping through the bandaged incision.

LINGUISTIC LEGACY

Adherents to the humour theory thought the liver, and not the heart, was the seat of courage. The heart was believed to be the home of learning, hence to 'learn something off by heart'. The liver was thought to house a reservoir of yellow bile, which supposedly deserted the organ in times of stress, hence the cowardly are sometimes said to be either 'lily-livered' or 'yellow-bellied'.

GREY MATTERS

Hippocratic medicine highlighted the role the brain played in influencing a person's emotions, an idea previously unfamiliar within Ancient Greece, when it was believed the heart was responsible for a person's mental functions. Humourists believed mental health was dictated by the humoural balance, which provided the foundations for

A look says it all: the faces of the four humours. (Clockwise from top) phlegmatic, sanguine, melancholic and choleric

modern psychology. To 'temper' something is to mix or blend, and the right blend of the humours was believed to engender a good 'temperament'. To the proto-psychologist there were four main temperaments, each dictated by the four cardinal humours. The sanguine were believed to be impulsive and extrovert yet also lofty and would take delight in being hurtful to those near to them; the choleric were aggressive control freaks who did well in the world of politics; the phlegmatic were passive-aggressive types who made good followers; and the melancholic, well, they spoke for themselves. While these beliefs were very broad, this was the start of a scientific understanding of what went on in the minds of those in a distracted or depressed state.

The alternative practice of homeopathy also has its roots in the humour theory. Distressed at the bloodletting, leeching, purging and other undignified and potentially terminal procedures that were being carried out, the German-born physician Christian Samuel Friedrich Hahnemann (1755–1843) investigated alternative procedures. Rather than letting fluids out of the body, Hahnemann pondered, why not impose a balance in the humours by more subtle means? He published his ideas in *The Organon of Homeopathic Medicine* (1810).

Doubtless influenced by the principles of inoculation and vaccination, as developed by the general practitioner Edward Jenner (1749–1823), Hahnemann focused on the concept of fighting like with like. He suggested that treatment of a medical condition should consist of administering a very dilute dose of substances that, in larger doses, would produce the symptoms that had initially ailed the patient.

Interestingly, in Greek, *homeo* translates as 'the same', and *pathos* means 'suffering'. Unfortunately Hahnemann took his idea to an extreme, and began speaking of magical dilutions that, when shaken by the practitioner, would release 'immaterial and spiritual power'. He believed that with a simple tap of the bottled dilution on the heel of the hand, the practitioner could 'double the dilution' of the medicine, that is half the strength yet again, and thus greatly increase its curative power.

Aside from Hahnemann's somewhat illogical notion that the weaker the dose, the more powerful the cure, homeopaths also believed that water possessed a memory of anything it might come into contact with; so a substance diluted to the point of virtual non-existence would be the most powerful of all. Many modern homoeopathists still believe this to be true, but let us hope they have been misled, otherwise our planet's water, which has been subjected to all manner of chemicals and faecal matter over the centuries, would surely be the most toxic substance known to man. Despite the far-out nature of some of Hahnemann's theories, he deserves to be held in fond memory for having saved many a patient from being drained of blood at a time when they needed it most.

As for Hahnemann's principles of dilution: I have conducted my own tests with whisky and, with my hand on my heart, I can affirm that one drop of single malt in a pint of water will not blow you out of your chair. It is far more effective neat.

Bibliography

50 Great Myths of Popular Psychology by Scott O. Lilienfield, Jay Lynn, John Ruscio and Barry L. Beyerstein (Wiley-Blackwell, 2010)
Bad Astronomy by Philip C. Plait (John Wiley & Sons, 2002)
Bad Medicine by Christopher Wanjek (John Wiley & Sons, 2002)
Bad Science by Ben Goldacre (Harper Perennial, 2009)
Boffinology by Justin Pollard (John Murray Publishers, 2010)
Elephants on Acid and Other Bizarre Experiments by Alex Boese (Pan Books, 2009)
Eureka! by Adrian Berry (Harrap Books, 1989)
The Greatest Benefit to Mankind by Roy Porter (HarperCollins, 1997)
The History of Medicine: A Very Short Introduction by William Bynum (OUP, 2008)
The Mad Science Book by Reto U. Schneider (Quercus, 2008)
The Skeptic's Dictionary by Robert Todd Carroll (John Wiley & Sons, 2003)
Trick or Treatment? by Simon Singh and Edzard Ernst (Corgi Books, 2009)
Science and the Practice of Medicine in the Nineteenth Century by William F. Bynum (CUP, 1994)
Science Was Wrong by Stanton T. Friedman and Kathleen Marden (Career Press, 2010)

Picture Acknowledgements

Pages 27, 31 www.karenswhimsy.com/public-domain-images; 50 Mary Evans Picture Library/Interfoto Agentur; 51 © Science Museum/Science & Society Picture Library (all rights reserved); 59 Library of Congress (LC-DIG-ppmsca-27955); 65 © The Art Archive/Alamy; 78, 115 www.clipart.com; 81 Walter Daran/Time & Life Pictures/Getty Images; 88, 138 Mary Evans Picture Library; 104 Courtesy of Institute for Nearly Genuine Research, www.bonkersinstitute.org; 110 Interfoto/ Sammlung Rauch/Mary Evans Picture Library; 114 Miles Kelly/fotoLibra; 140 Roberto Castillo/www.shutterstock.com: 144 lynea/www.shutterstock.com

Index

(page numbers in italics refer to illustrations)